Sophie's Diary

Second Edition

The first edition of *Sophie's Diary* was self-published by Dora Musielak.

Second Edition
© *2012 by the Mathematical Association of America, Inc.*

Library of Congress Catalog Card Number 2012934877

Print edition ISBN: 978-0-88385-577-5
Electronic edition ISBN: 978-1-61444-510-4

Printed in the United States of America

Current Printing (last digit):
10 9 8 7 6 5 4 3 2 1

Sophie's Diary

Second Edition

Dora Musielak

Published and Distributed by
The Mathematical Association of America

SPECTRUM SERIES

The Spectrum Series of the Mathematical Association of America was so named to reflect its purpose: to publish a broad range of books including biographies, accessible expositions of old or new mathematical ideas, reprints and revisions of excellent out-of-print books, popular works, and other monographs of high interest that will appeal to a broad range of readers, including students and teachers of mathematics, mathematical amateurs, and researchers.

MAA Service Center
P.O. Box 91112
Washington, DC 20090-1112
800-331-1622   FAX 301-206-9789

Contents

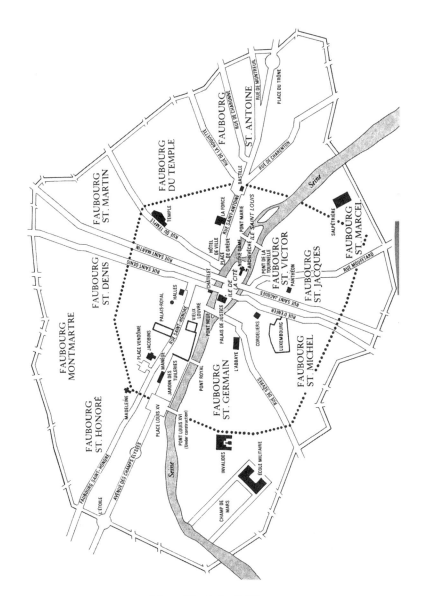

Map of Paris in 1789.

Paris, France
1789

A long time ago, in a far away land, a sage woman raised three boys to be honest and smart. When her husband died, he bequeathed her one milking cow, and thirty-five more to be divided among their three sons. According to the wishes of the father, the oldest was to receive half the number of cows, the second would get a third of the total, and the youngest would receive a ninth of all.

The brothers began to quarrel amongst themselves, as they did not know how to divide the herd. They attempted different ways to divide it, and the more they tried, the angrier they became. Every time one of the brothers proposed a way to divide the thirty-five animals, the other two would shout that it was unfair.

None of the suggested divisions was acceptable to all. When the mother saw them squabbling, she asked why. The oldest son replied: "Half of thirty-five is seventeen and one-half, yet if the third and the ninth of thirty-five are not exact quantities either, how could we divide the thirty-five cows amongst the three?" The mother smiled, saying the division was simple. She promised to give each their share of the herd and assured her sons it would be fair.

Do you know how the wise woman divided the herd?

1
Awakening

Wednesday — April 1, 1789

How can I divide 35 in a half, a third and a ninth, and get an even number? It seems impossible! Yet I know there must be a way....

My name is Sophie. Today is my thirteenth birthday and my parents surprised me with wonderful gifts. My father gave me a lap desk of glossy mahogany wood. It has a compartment to keep paper and pens, and a secret drawer that can only be opened by pressing a hidden tab underneath. When I unlocked it, I found a mathematical riddle and a note from Papa, challenging me to solve it. He gives me a year, but I hope to find the answer sooner pg. 43 than that. Oh, I am so excited!

My mother gave me a porcelain inkwell, a stack of linen paper, sealing wax, a miniature burner, six bottles of ink in different colors, and a set of quills. These are the best birthday presents ever! I promised my parents to write often to improve my penmanship. Maman says writing neatly and correctly is very important for a refined young lady. This gave me the idea of writing a *journal intime*, a diary where I can record my innermost thoughts and feelings.

I love my father. He taught me arithmetic when I was five. I still remember my excitement when Papa taught me the numbers. I used to think that the arithmetic operations were magic tricks to produce new numbers. Just for fun, I spent hours adding numbers to create others. Now I help Papa

1

with the more complicated calculations for his business transactions. He still calls me *ma petite élève*.

I have two sisters, Angélique-Ambroise and Marie-Madeleine. Angélique is ten years old, a pretty and vivacious girl who talks incessantly, exactly the opposite of me. Madeleine is almost nineteen years old. She resembles our mother and, like her, she is romantic and sweet. I am not like either of them. My sisters like to dress in beautiful gowns, style their hair, and sing and dance. I don't like any of that.

Angélique calls me "peculiar." She teases me all the time because I like to be alone and would rather read books. I do not care whether my hair is curled, or if my dresses are fashionable. I do not think about these things the way she does. But I am fond of her, just as I am of my mother and Madeleine. I especially love my father with all my heart; *mon cher père*, he understands me better than anyone.

We live in a townhouse on rue Saint-Denis not far from the Seine River. It is one of the most important avenues in Paris. Kings and princesses enter the city through Saint-Denis. This neighborhood is lively. From my bedroom window I see the imposing buildings of the Conciergerie and the Palais de Justice. I can hear the bustle of the people on the Pont-au-Change, the bridge over the river that we cross going to Mass at the cathedral Notre-Dame de Paris.

The church bells are tolling the eleventh hour. I must stop writing now and go to bed. But I still feel giddy, thrilled with happiness and tenderness towards my parents for giving me these wonderful gifts for writing. From now on, I shall use ink and paper to express my own ideas and thoughts, just as my parents suggest.

Sunday — April 5, 1789

What a lovely day! A bit cold but the sun broke through the clouds and the last traces of ice lingering on the rooftops began to melt. Paris awakes from the long slumber of winter, so harsh this year. In January, it was so cold that the water in the house froze. We were sick with sore throats and coughing. My mother was worse, as she succumbed to pneumonia. Thankfully, she recovered and now she looks radiant.

My mother is very traditional, an old-fashioned lady who insists that my sisters and I learn the graces of society. She teaches us good manners, just as her mother taught her. She believes that young women must learn domestic arts to become good wives. I won't, for I will never get married! Madeleine, however, hopes to wed Monsieur Lherbette very soon. I like him. Sometimes he converses with me in the foyer while waiting for my sister. He is always polite and asks me questions about the books I am reading.

Our family life is simple. We begin every day with prayer, and then my father retires to his library to review his business accounts. We dress and get ready for the day. After breakfast my father goes to work, and my sisters and I join Mother in her sitting room. We spend the morning sewing, knitting, and learning needlework. When the weather is nice, we go for a stroll in the gardens. Angélique plays with other children while I read next to Maman, sitting on a bench by my favorite tree. Once a week, Monsieur Chevalier comes to give me piano lessons. I find these lessons tiresome. Not that I do not like music; on the contrary, I love it, but I would rather listen to others play. I have no talent, especially compared with Madeleine. Even Angélique plays the piano much better than I do.

Papa comes home to eat lunch at one o'clock. In the afternoon my mother instructs me in penmanship. This is a lesson I enjoy because Maman chooses passages from important books that I copy to help me learn the correct manner of writing. Every Friday after dinner, my mother reads full chapters of historical novels or scenes from Greek plays and then she queries us about them. Angélique and Madeleine recite love poems. When Papa is home, he talks about philosophy and history, his favorite subjects. After supper, my mother and Madeleine play cards with Papa. Every Tuesday night Papa and his colleagues meet in his study to discuss politics. I read or play Chinese checkers by myself, but if Papa is not busy he plays chess with me. He is very good, but I usually win the match.

I dislike embroidering, but Maman will not excuse my ineptness and makes me try harder. I'm so clumsy with handicrafts and can never finish even the simplest design. At night, Millie tiptoes into my bedroom and finishes the embroidery for me. I hope Maman does not find out or she'll be upset with me and reprimand her. Millie's job is to help Madame Morrell with the household chores, run errands, and look after my little sister. Monsieur and Madame Morrell are Millie's parents. The three have worked for my family for many years. Monsieur Morrell is Papa's valet and coachman, and Mme Morrell cooks, runs the household, and assists Maman with her coiffure.

My favorite place is my father's library, where I read every day. I also like to be there because I can listen to the conversations Papa has with his friends. They talk about politics and that usually leads to discussions of philosophy, literature, and other interesting topics. The best part is that the gentlemen don't seem to mind that I am there, sitting quietly in a corner. Or perhaps they don't notice me.

My father is very knowledgeable about the financial state of France. He is worried about the shortage of goods and the increase in taxes, problems affecting mostly the poorest citizens. Papa says that is why many workers are protesting and demanding more social equality. Some demonstrations have become violent. The people blame King Louis-Auguste XVI for the failing economy. Papa believes that the peasants and poor workers are taxed too heavily. He supports the working class and their demands for a fair share of privileges but admits it may be difficult to change a society like ours. As it is, aristocrats and church bishops possess the wealth and enjoy all the social privileges. Maman defends the king, but she can't excuse the social injustices.

Oh, I hear the clock chime the midnight hour. I must go to sleep now or I won't be up on time for morning prayers.

Friday — April 10, 1789

Late last night I visited my father's library looking for something to read and found a rather fascinating book. It is about the history of mathematics, describing how mathematics evolved since the beginning of recorded history. I was so engrossed in reading that I didn't go to sleep until the bells rang midnight. The author, a mathematician named Jean-Étienne Montucla, devotes the first chapters to the development of mathematics in antiquity.

The part that I found most fascinating was the story of Archimedes, an amazing scholar who lived in a Greek colony called Syracuse hundreds of years ago. Archimedes was an exceptional man. The author describes Archimedes as absentminded, just like me. Archimedes would become so involved in his study that he paid no attention to what was happening around him. He'd draw geometric figures on any available surface, including the dust on the ground or in the ashes from extinguished fire. Archimedes solved a very interesting problem that goes like this. One day the king of Syracuse gave an artisan a certain amount of gold to make him a crown. The finished tiara was beautiful, but for some reason the monarch suspected that the craftsman had stolen some of the gold and replaced it with silver to make up the required amount needed to craft the crown. The king could not prove the artisan's deception, so he asked the wise scholar for advice.

Archimedes had to find a rational method to confirm or refute the king's suspicion, without destroying the crown's exquisite workmanship. One day, when the sage man entered his bathtub, he observed that the amount of water that overflowed the tub was proportional to the weight of his body submerged in the water. That gave him the idea to solve the problem of the king's crown. Archimedes was so excited that he ran naked through the streets of Syracuse shouting like a lunatic *j'ai trouvé, j'ai trouvé*. The author used the Greek word *Eureka*, an interjection used to celebrate a discovery, which means, "I have found (it)! I have found (it)!" What Archimedes had discovered was the Law of Buoyancy, as described in his book *On Floating Bodies*. This law states that a floating object always pushes aside an amount of water equal to the weight of the object. Archimedes also discovered that a nonfloating object—one that sank—pushed aside an amount of water equal to the object's volume. The great scholar went on to prove that the crown was indeed made with two different metals.

Archimedes died tragically while doing mathematics. It happened the year 541 of Rome, and 212 years before the Christian era. The good scholar was so absorbed in his studies, that he was unaware of the Romans invading Syracuse. As was his custom, that day Archimedes was calculating something, drawing geometrical figures in the dust when a Roman soldier stepped on them. Archimedes was very upset and told him, "Do not disturb my circles!" The soldier was so enraged by the old man's outburst that he pulled out his sword and killed him.

After reading this part of the book, I could not sleep. I kept imagining Archimedes, a person so consumed by a mathematical problem that he did not notice what was going on around him, not even a foreign invasion. A strange feeling overcame me, a need to experience the same kind of passion that Archimedes must have felt.

I understood my desire to know more of this science. Mathematics is not just about the simple calculations that ordinary merchants perform every day. Oh, I am happy to know basic arithmetic, to add and multiply numbers, but it is insufficient. No, it is not enough that I know the simple operations that Papa taught me. I want to learn much more to become a great mathematician, just like Archimedes.

There are many concepts that I wish to understand, like the number pi, denoted π, which is prominent in the story of Archimedes. I need to know what π means and why it is so important. Reading a book of mathematics, full of symbols and equations, is like reading a book written in another language. I want to speak the language of mathematics!

How I wish I had someone to teach me. Antoine-August, the son of one of Papa's friends, is studying mathematics with a tutor. This morning I asked Maman if I could have a teacher too, but she stopped me, looking at me as if I was insane. Frowning, Maman stated that girls do not study mathematics and asked me to expunge those thoughts from my mind. Why? I am a girl and I like mathematics very much. Why should I stop thinking about it?

Wednesday — *April 15, 1789*

Now I know what the symbol π means. First, π is a letter in the Greek alphabet used by mathematicians to represent an important number. As a number, π represents the ratio of the circumference to the diameter of the circle. It is such a unique number that there is no fraction equal to it, and its value is the same for all circles, regardless of their size.

To prove it, I could draw a circle and measure its circumference with a piece of string, which I would place on the line that represents the circle, exactly once around. The length of the string should correspond to the circumference, C, of the circle. I would measure the diameter, D, of the circle, which is the length from any point on the circle straight to its center to another point on the opposite side of the circle. Then I would divide the circumference of the circle by its diameter. This ratio C/D is equal to π. Interestingly, if I draw smaller or bigger circles, the result is the same: C/D is *always approximately* equal to 3.141592. In other words, π is a constant with the same value for all circles.

But if I could measure and divide perfectly, could I get an exact value of π? Based on what I know, the answer is no! The number π is *not* an exact ratio. The approximated value of π has been known for thousands of years, but its value was obtained by measurement, therefore it was not very accurate. One day, Archimedes proposed a mathematical procedure that used geometry instead of direct measurement. In his book *The Measurement of a Circle*, Archimedes described how he took a polygon with 96 sides and inscribed a circle inside the polygon, and after performing a geometric analysis he estimated the value of π as 22/7, stating that it is a number between $3\frac{10}{71}$ and $3\frac{1}{7}$.

Now mathematicians write $\pi = 3.141592\ldots$. The dots after the last digit mean that the exact value of π is unknown. One can keep on refining the calculation, thus getting more digits on the right side of the decimal point, but although mathematicians have calculated many digits, so far nobody knows how many more there are.

I've read two full chapters of the *History of Mathematics* and learned about another ancient mathematician named Pythagoras of Samos. He was a Greek philosopher who also studied astronomy and developed music theory. Pythagoras moved to Croton, a city in southern Italy (a Greek colony) and started a religious and philosophical school. This great man had many followers called the Pythagoreans, scholars who also studied mathematics; after the master died, they continued the schools for many more years. Pythagoras believed that most things could be understood through numbers. He and his followers studied numbers and gave some of them a special significance. For example, the Pythagoreans thought the number 10 was the very best number: it contained in itself the first four integers: $1 + 2 + 3 + 4 = 10$. They also defined "perfect numbers." According to Pythagoras, numerical perfection depended on a number's divisors. A perfect number is a positive integer that is the sum of its positive divisors, that is, the sum of the proper divisors. For example, the number 6: its proper divisors are 1, 2, and 3, and consequently 6 is a perfect number because $1 + 2 + 3 = 6$. The next perfect number is 28, because $1 + 2 + 4 + 7 + 14 = 28$. I have to find the next perfect number!

I could not find much about the life of Pythagoras, but I learned that in addition to studying the properties of numbers, the ancient scholar is best known today for a theorem in geometry. Although Pythagoras's theorem was already known to the people of his time, the historian states that Pythagoras may have been the first mathematician to prove it. His work with numbers and the famous theorem led to developments of new mathematical concepts. The author states that much of the mathematics of today is due to Pythagoras and his followers. That's why I intend to investigate more about this great man and the numbers he studied.

Sunday – April 26, 1789

This afternoon Angélique made fun of me and called me *gauche*. I did not say anything, but her incessant teasing hurts me. We came from church and were having Sunday lunch; everybody was in a fine mood. Then Mother mentioned the invitation from Madame LeBlanc to attend a dinner party next week. I do not like social gatherings where everybody talks and laughs, telling each other silly stories. I never know what to say, and I feel awkward and out of place. It doesn't help that Angélique makes fun of my shyness. The worst part is that, like it or not, I have to go to Mme LeBlanc's soirée.

Mother thinks it is important for my social development, that it will teach me to behave like a lady.

I would rather stay home and read books, to learn about extraordinary people like Pythagoras. Numbers fascinated him, thinking that they possessed mystical powers. Amazingly, he found relationships between numbers and things that at first appear unrelated to mathematics, like music! Pythagoras discovered the relationship between the length of a taut string of a harp and the sound it produces when the string is plucked. He experimented with different lengths and found that if the string is shortened to half of its original measurement, then the tone produced is one octave higher than the original tone. That's how this great scholar discovered musical intervals. Pythagoras influenced the development of mathematics to this day. He founded a school for men and women to study mathematics and science. What I read implies that the wife of Pythagoras was also a mathematician, but her name is not given. Why is that? Does it mean her accomplishments were not as important as her husband's? Or that she was shy like me and didn't like to show off how smart she was? Nobody knows. After all, these events happened centuries ago!

Tuesday — April 28, 1789

I always thought Paris was the richest city in Europe, but now I am not so sure. Many people are poor and wander the streets begging. And food is scarce. Millie told Maman this morning that people started a fight when the baker ran out of baguettes. To ensure we get enough bread, Millie goes to the bakery very early and waits in a long line for hours. The people that come late don't get any and the poorest cannot afford it. According to Papa, laborers earn twenty-five sous a day, and now a loaf of bread costs fourteen or fifteen sous! Millie added that some workers were exchanging their shirts for a loaf of bread and, in one case, a woman even tried to exchange her corset.

A demonstration erupted at the market yesterday. Millie told us how it happened, that it began with a speech against social injustice. Then a man began to shout, inciting the shopkeepers to revolt, and soon several others followed. We heard it, as the people took to the streets. The mob chanted, "Death to the rich! Death to the aristocrats!" and were growing more rowdy by the minute. The violence of the protests is escalating. It scares me!

Later in the afternoon, another riot broke out in the faubourg Saint-Antoine. The angry workers, acting on rumors of wage cuts, destroyed the wallpaper

factory of Monsieur Réveillon, a business friend of Papa. Many peo
killed when the government troops opened fire at the mutineers. T
ers, afraid of losing their jobs or not earning enough to afford bread, started
with a protest. One thing led to another, and the protest boiled over into violence. The fire also consumed the family's house. Their great library, with
hundreds of valuable books, was burned to ashes and now there is nothing left. The only good thing is that M. Réveillon and his family escaped
unharmed.

I continue reading. The *History of Mathematics* is one of those precious
gems in Papa's library that are too hard to put down. I finished another
fascinating chapter, full of anecdotes about ancient scholars. I read in another book about Hypatia of Alexandria, the first woman scholar of considerable merit. Hypatia was the daughter of the mathematician Theon, who
lived around 400 A.D. This remarkable woman was the head of a school
in Alexandria where she taught the ideas of the Greek philosopher Plato.
Scholars of that era esteemed Hypatia for her intelligence and vast knowledge. Hypatia helped her father write commentaries on important mathematical books, such as Euclid's *Elements*. Hypatia's greatest achievement
was her commentaries on the *Arithmetica* of Diophantus, one of the most
important books for mathematicians. Diophantus of Alexandria was an ancient Greek mathematician who lived sometime before Christ.

Hypatia died most tragically, massacred by people who felt threatened by
her intelligence and knowledge. A rumor started that Hypatia was responsible for a conflict that arose between two important spiritual leaders. It is not
clear why she was blamed and why some men disliked Hypatia so much.
One day the men seized her on the street, beat her to death, dragged her
body, mutilated her flesh with sharp tiles, and burned her remains.

Poor Hypatia; what a horrible way to die! How hideous such an assassi-
nation by the barbarian mob; how cruel those people were! I tremble just
imagining her nightmare.

Saturday — May 2, 1789

My father was elected deputy of the Constituent Assembly. In this capacity, he will represent the citizens of the Third Estate before the court. Next
week he will go to Versailles to attend a meeting with the king and discuss the economic problems. Father knows of the country's huge debt and
is aware of its effects on society. In fact, he suggests several social reforms

to restore balance. Many people blame King Louis XVI and Queen Marie-Antoinette for their economic problems. But the debt was already immense when they ascended the throne. Regardless of who is at fault, the tax increase is creating serious social and economic conflicts.

My father acknowledges that taxes are not distributed equally among the citizens of France. That is the reason why the workers and peasants are so unhappy. It's an unfair situation for them since the clergy and aristocrats, who represent the First and Second Estates, are exempt from taxation. That leaves only the Third Estate—the rest of the people—to pay the debt. Any reasonable person would see an obvious solution: force the nobles and the clergy to pay their share of taxes. A change like this one would strengthen the ideals of equality in our society. Father agrees, but he says the privileged aristocrats are fiercely opposed, and they will do anything to avoid losing their wealth and property.

I never thought about it before, but now it all becomes clear. I begin to understand what the French philosophers of the Enlightenment meant about social inequality. Why are there social classes? When did all this begin? Long ago, when the primitive societies formed, someone assumed power; but how? Who dictated that one person would control the lives of many others? Right now, Louis XVI rules France because he inherited the throne from his grandfather. I can name the kings of France in the recent past, but I cannot tell who the first king was. How did the monarchies of the world begin? It is not fair that, under this social structure, only a few individuals have power and wealth, and the remaining citizens have none. My father has stated many times that a monarchy is unnecessary. He thinks it is possible for a country to exist without a king. Can France become a republic ruled by the people?

Monday — May 4, 1789

Today was a historical day for France. It began with a great parade through the streets of Paris to inaugurate the dialogue between the Estates General and Louis XVI to solve the economic problems. This exchange of ideas is the first of its kind in the history of France. The king recognizes the needs of his subjects and is willing to listen to their demands. We, like all Parisians, went to watch the parade. But it was more meaningful for us, as Father was in the procession with the other deputies of the Third Estate.

King Louis XVI and Marie-Antoinette led the solemn march. Princes and princesses, wearing the most beautiful attire, followed the royal family.

The ladies wore stunning brocade and taffeta gowns, and their hair was piled high, adorned with colorful feathers. They were resplendent, with their jeweled tiaras shimmering against the sun. The queen seemed a bit forlorn. Maybe she still mourns the baby she lost two years ago, or maybe she is worried for the Dauphin of France, her sickly seven-year-old son. But she walked with grace and dignity and smiled to her subjects. The bishops and other leaders of the Church of France marched behind the aristocrats, wearing their richest robes of silk and lace. The royal falconers strode with hooded birds attached to their wrists.

As the parade made its way, the people lining the streets shouted, "*Vive le roi, vive la reine!*" A few men bellowed, "Long live le Duc d'Orleans!" to acknowledge the king's cousin who ardently supports the ideals for social equality.

And last, the deputies of the Third Estate walked somberly behind, all dressed in simple black outfits that were in stark contrast to the rich costumes of the clergy and the nobility. Even without such regalia Father looked so handsome! With his virile walk and stature, Papa was a commanding presence among the six hundred deputies. Angélique waved at him excitedly and yelled, "*Vive le tiers état!*" I doubt he saw us among the immense crowd.

The parade culminated with a high Mass at the Church of Saint-Louis but we didn't attend. We returned home, tired and hungry, but excited and so proud of Papa. As a deputy of the Third Estate he will get an opportunity to express his ideas for economic reform and equality.

Sunday — May 10, 1789

Some people believe that Babylonian mathematicians knew the Pythagorean Theorem hundreds of years before Pythagoras established it in the form we know it. The Pythagorean Theorem is a mathematical statement of great simplicity. It says that in a right triangle, the square of the hypotenuse is equal to the sum of the squares of the legs. I can write the theorem using an algebraic equation: $x^2 + y^2 = z^2$, where x and y are the sides of the right triangle, and z is the hypotenuse.

It is a rather simple equation, yet it is very powerful. I could use the Pythagorean Theorem to calculate areas, lengths, and heights. For example, suppose I have two points on a plane, and I want to know how far apart they are. I can either measure it directly, or I can measure the distance along the

x-axis, measure the distance along the y-axis, and then use the Pythagorean Theorem to find the distance. In many situations it is easier to measure the x and y distances than the total distance between two points.

For example, suppose that I have a right triangle for which I know the length of the two short sides, say $x = 7$ and $y = 3$. Then I can calculate the value of the longest side of the triangle, or hypotenuse, by applying the Pythagorean Theorem. The square of the hypotenuse z is equal to the sum of the squares of the two sides:

$$z^2 = 7^2 + 3^2$$
$$z^2 = 49 + 9$$
$$z^2 = 58$$
$$z = \text{ square root of } 58.$$

The square root of 58 is $7.61\ldots$, a number that has a decimal part with many digits.

Now, if $x = 3$ and $y = 4$, I get:

$$z^2 = 3^2 + 4^2$$
$$z^2 = 9 + 16$$
$$z^2 = 25$$
$$z = 5.$$

In this case the Pythagorean Theorem yields perfect squares, $5^2 = 3^2 + 4^2$. A perfect square is a number whose square root is a whole number.

The Pythagorean equation leads me to two types of numbers: whole numbers and fractional numbers. Now, what about negative numbers? Is it possible to get negative numbers? If I multiply a negative number by itself, would it yield a positive number?

I am sure there is much more to learn about this theorem. But it has to wait until tomorrow. The church bells tolled the eleventh hour a while ago. It must be midnight already!

Saturday — May 16, 1789

I am tired. We spent almost all day at the Palais-Royal. This is the most popular place in Paris, especially during the weekend. It was crowded, with people strolling in the congested archways. Women wearing the most extravagant hats strutted along, proudly displaying their fashionable dresses.

We went to a café and found a table under the trees in the central court-yard. Maman ordered Austrian pastries and, while we enjoyed our delicious dessert, we watched a troupe of actors made up like marionettes. Gypsy women wearing colorful skirts also danced, and some were going around reading people's fortunes.

A young man suddenly jumped on a table nearby and began a speech, inciting citizens to fight for equality and justice. Some people stopped to listen, but many more ignored the eloquent orator. A few men distributed pamphlets with invitations to join groups to fight injustice. The message was clear: France needs to start a revolution.

At noon we watched the solar cannon. This is another popular diversion that attracts many onlookers. My sister Angélique was the first in line to see the cannon firing at midday, as she said, "to salute the sun." But she was disappointed that the cannon did not detonate again. I had to explain to her that it can shoot only once a day, when the rays of the sun strike the lens at the noon hour. The lens acts as a burning glass, which precisely focuses on the touch-hole of the small cannon, and so it fires automatically, just once a day. I'm not sure Angélique understands that.

I have my own amusement with mathematics. I've been thinking about the number π. I know that, according to Archimedes, $22/7$ is a good approximation to its value, and $355/113$ is a much better one. But there is no exact number, all I can write is $\pi = 3.141592\ldots$. Most likely the number expansion on the right side of the decimal point is infinite. Maybe it is infinite because the circumference of the circle is an infinite number itself. This is what I think: since π is equal to the ratio of the circumference to the diameter of a circle, C/D, and the circumference C is the length of the line that represents the circle, and this length does not have a beginning or an end, I conclude that dividing something infinite must give an infinite value.

In addition to its relation to the circle, why else is this number so important? Perhaps π is the solution to an equation that describes the universe. Especially, if the universe were round! Oh, I am sure that learning mathematics can help me answer these and many other questions that burst in my mind.

Wednesday — May 20, 1789

A deafening thunderstorm woke me up a few minutes ago. As the lightning parted the darkness of the night, I jumped out of bed. The room became fully illuminated as if a thousand candles were lit at once; but just as fast, it became pitch dark. The roaring continued and rain was falling hard. It must be about three o'clock in the morning!

Few things scare me, but thunderbolts terrify me. I've never told this to anyone; who would understand? During a lightning storm I tremble like a scared child. I wish I could find refuge in my mother's arms, as I imagine the warmth of her embrace would soothe my fear; the sound of her beating heart would muffle the frightful sounds. But I am not a child anymore. I must confront my fears and be strong. I will just sit here and wait until the storm ends.

I know my fear of lightning is irrational. Irrational as in "not rational" or illogical. This reminds me of something. In my study of numbers, I found rational and irrational numbers, but in mathematics, "irrational" does not mean illogical. In fact, irrationals are some of the most amazingly beautiful numbers, although they seem to be very rare. Irrational numbers are real numbers that cannot be represented as fractions. Irrational numbers result from dividing two real numbers, but they are not perfect ratios like $1/2$, $12/3$, $24/4$, which are called rational numbers.

So, if the ratio of two real numbers results in a number with a decimal part that goes on and on without end, and the digits in the decimal part are randomly distributed, then the number is irrational. For example, π, which is approximately equal to $3.141592...$, is an irrational number because its value results from dividing two real numbers (the circumference and the diameter) and the digits on the right side of the decimal point go on without settling into a repeating pattern. How many digits? I do not know, but it's probably an infinite number.

Hundreds of years ago, Pythagoras discovered that harmony in musical instruments is determined by the ratio of the lengths of the strings. He also thought that the planets orbit in harmony to ratios. Pythagoras considered ratios as the mystical concept that the Divine Mathematician used to create the universe and everything in it. Pythagoras rejected any number that did not conform to this ideal. Thus, the Pythagorean mathematicians that followed his teachings also overlooked irrational numbers.

But irrational numbers do exist; an example is the square root of 2. This number is a fraction with a non-repeating decimal that goes on without end. This non-ratio later became known as an irrational number. The Pythagorean philosopher Hippasus used geometric methods to demonstrate the irrationality of $\sqrt{2}$. There is a legend that tells how Hippasus was thrown overboard off a ship for studying irrational numbers and proving that the square root of

2 is irrational. When Hippasus announced his discovery, his outraged colleagues threw him overboard for going against the teachings of Pythagoras.

I don't think it really happened. All I know is that Pythagorean mathematicians did not work with irrational numbers. This is the only disappointing fact about the great mathematician of antiquity.

Thursday — May 28, 1789

I think about numbers all the time. Numbers are not just tools for adding or subtracting, as that would make mathematics too simple and uninteresting. I think about all types of numbers such as integers, rational and irrational numbers, primes, and perfect numbers. I want to learn what makes these numbers different and how to use them.

Integers are numbers that do not have a decimal part like 1, 13, 27, and so on. When I write 1, 2, 3, I say that these are consecutive integers. In general, consecutive integer numbers are integers n_1 and n_2, such that $n_2 - n_1 = 1$; i.e., n_2 follows immediately after n_1. So, given two consecutive numbers, one must be even and the other must be odd. Since the product of an even number and an odd number is always even, the product of two consecutive numbers, and, in fact, of any number of consecutive numbers, is always even.

An integer of the form $n = 2k$, where k is an integer, results in even numbers, such as, $-4, -2, 0, 2, 4, 6, 8, 10, \ldots$. The product of an even number and an odd number is always even, as can be seen by writing $(2k)(2m + 1) = 2[k(2m + 1)]$, which is divisible by 2, and hence is even.

A number that can be expressed as a fraction p/q, where p and q are integers and $q \neq 0$, is called a rational number with numerator p and denominator q. When the fraction is divided out it becomes a number with a terminating or repeating decimal, such as $1/2 = 0.5$, and $5/3 = 1.666666\ldots$

Fractional numbers that are not rational are called irrational numbers. An irrational number cannot be expressed as a fraction p/q. In decimal form, irrational numbers do not repeat in a pattern or terminate; they go on and on, such as $\pi = 3.141592\ldots$ and $\sqrt{2} = 1.414213\ldots$

And then there are primes, the most intriguing and fascinating among all numbers. A prime number is a positive integer p that has no positive integer divisors other than 1 and p itself. More concisely, a prime number p is a positive integer having exactly one positive divisor other than 1. For example, the only divisors of 13 are 1 and 13, making 13 a prime number, while the number 24 is not because 24 has divisors 1, 2, 3, 4, 6, 8, 12, and

24, corresponding to the factorization ($24 = 2^3 \cdot 3$). Although the number 1 was considered a prime number, it is not. Thus, *excluding* 1, the first few primes are $2, 3, 5, 7, 11, 13, 17, 19, 23, 29, 31, 37\dots$.

This is how I see which numbers are primes. First, I write all the positive numbers:

1	2	3	4	5	6	7	8	9	10
11	12	13	14	15	16	17	18	19	20
21	22	23	24	25	26	27	28	29	30
31	32	33	34	35	36	37	38	39	40
41	42	43	44	45	46	47	48	49	50
51	52	53	54	55	56	57	58	59	60
61	62	63	64	65	66	67	68	etc	etc

Then I take away the numbers that are multiples of 2, then those that are multiples of 3, of 4, 5, 6, and so on. The numbers that remain are the prime numbers!

	2	3		5		7			
11		13				17		19	
		23						29	
31						37			
41		43				47			
		53						59	
61						67			

Interestingly, there is only one even prime number. Does this mean that, with the exception of 2, all the primes are odd? How can I check this without having to calculate one by one? There may be many prime numbers, infinitely many perhaps, so I cannot check one by one. Surely there must be infinitely many primes since there must be infinitely many integer numbers. If I can add a 1 to any integer number, then no matter how large my last integer is, I can always make it larger by adding one, which means the number of primes continues increasing, without end. And prime numbers are integers too, thus there must be an infinite number of primes as well. How can I prove it?

Thursday — June 4, 1789

France is in mourning. Today after ten o'clock, the bells in every church in Paris tolled to announce that the Dauphin of France passed away. The young son of King Louis and Queen Marie-Antoinette died of tuberculosis. Mother sent us to say a prayer for the soul of the little boy.

After meditating for a while, my mind began to wander. I thought about Pythagoras and wondered why such a brilliant mathematician rejected ratios that are not perfect. Why did Pythagoras not accept numbers like the square root of 2? But then an idea burst in my head. As I gazed up, a shadow from a window formed a right triangle on the wall. The two sides of the triangular shape seemed to be the same size, so I arbitrarily made them each equal to 1. Then I asked, what is the length of the hypotenuse of such right triangle? To answer, I recalled Pythagoras's theorem: $x^2 + y^2 = z^2$.

In this case, $x = 1$ and $y = 1$, therefore $z^2 = 2$. This means that I have to find a number which, when multiplied by itself, gives me the number 2. The hypotenuse z is equal to the square root of 2; i.e., $z = \sqrt{2} = 1.414213\ldots$. Thus, $\sqrt{2}$ is an irrational number because it cannot be written as a ratio of integers p/q. Why did Pythagoras not see that, too?

Wednesday — June 10, 1789

This morning I announced my desire to be a mathematician. How I regret my childish outburst. How I wish I didn't say it! We were eating and, as usual, my sister was chattering away when, without thinking, I blurted it out. Mother's sweet smile froze on her lips, and then she shook her head but said nothing. Angélique swallowed fast and interjected, "But Sophie, mathematics is not for girls!" and she looked at me as if I were mad.

I was about to explain to her that women can also be mathematicians like Hypatia, but before I could utter a word my father turned to me and, patting my hand tenderly, said, "If Sophie wants to be a mathematician, then she *will be* a mathematician." I wanted to kiss him and hug him as I used to do when I was a little girl. But Mother's austere look kept me frozen in my seat. Surprisingly, Angélique did not comment further; she kept blabbering about her new dress.

After Father left the table my mother admonished me. She thinks it is unbecoming of a young lady to speak such nonsense and chastised me for "filling my head with such outrageous ideas." What is wrong with my wish to become a mathematician? I don't understand why Mother is against my desire to study something I like so much. She claims that mathematics is not part of my education. She forbade me to stay up at night, reading such "an unfeminine subject." Who decided mathematics is unfeminine? Why should mathematics be just for men? I don't accept it!

Still, I wish I didn't say anything at all. But there is no turning back now. Yes, I will become a mathematician somehow! Mother cannot prevent it. There is nothing else in the world that interests me more. I do not want to spend my life in front of the mirror curling and powdering my hair. I only want to study mathematics.

Monday — June 22, 1789

My mother gave me a stern lecture. She discovered that I was reading in bed past midnight, disobeying her orders. But that is not all. She walked into my bedroom when I was writing equations and I didn't have time to hide the notes in my drawer. I would have spent the entire night studying, but Mother came and forced me to bed. If I were writing a letter she would not have made such as fuss, but discovering I was working on mathematics made her very angry. She seized my notes, saying she'd burn them; I cried almost all night.

In the morning before breakfast, Mother scolded me again. I don't understand why my mother thinks that mathematics is an unfeminine subject, not proper for a girl. Why does my desire to study mathematics anger her so much? I wish Papa were here. He knows I like numbers; he will stand by me, I am sure of that. After all, I studied arithmetic under his tutelage.

My father is very busy with his work in Versailles and will not return until next week. The assembly with King Louis turned into an ugly affair due to the disputes and arguing of the representatives of the Three Estates. Monsieur Morrell came this afternoon to bring a message from Papa. He told Maman that the arguing became so heated that the deputies of the Third Estate declared themselves the National Assembly, and then they made a pledge not to separate until they give France a constitution.

Millie came to my bedroom before she extinguished the lamps in the hallway. She brought me a cup of hot chocolate and offered to help me finish the handkerchief my mother asked me to stitch. Millie knows that I would rather read a book than do these chores, so I accepted gladly. While Millie embroidered, I read the history of mathematics until I reached the end of the book. I did not notice that Millie left, so it must be late.

It's so easy to get lost in the pages of this book. There are unforgettable stories about the ancient mathematicians, each one pointing me in a different area of study. Reading the history of mathematics makes me realize how much I have yet to learn. For now, I snuff out my candle before Maman finds out I am still awake.

Thursday — June 25, 1789

Millie saved my notes! She found the pages that Maman threw away and hid them for me. I am so relieved! I thought I had to start all over again. What would I have done if my notes disappeared in the flames? Begin again. The numbers I wrote, the equations I solved, all that I would repeat again and again.

Mother is not here this evening. She is at the theater chaperoning Madeleine and her fiancé. This allows me a few hours to myself. I know Maman disapproves and would be terribly angry if she discovers I am back at my studies, but I am very curious about something that Antoine-August mentioned the other day. His tutor is teaching him *polynomials*. He did not say what a polynomial is, but I was intrigued and so I looked it up. I found that a polynomial is defined as any mathematical expression consisting of a sum of a number of terms. What does it mean exactly?

If I have any three terms, such as x, y, z, I could add them and write the sum: $x + y + z$. But that is meaningless, unless I add an equal sign and make it an equation, like $x + y + z = 0$. Now, if I use the same variable x raised to a power like x, x^2, or x^n, then I can define a polynomial as the sum of these terms:

$$x^2 + x^3 + x^4 + \cdots + x^n = 0.$$

Is this correct? Oh, how can I be certain? If a polynomial is a mathematical expression involving a series of powers in one or more variables multiplied by coefficients, then I can write a polynomial in one variable with constant coefficients as:

$$a_n x^n + \cdots + a_3 x^3 + a_2 x^2 + a_1 x + a_0 = 0.$$

The highest power in the polynomial is called its order. And it makes more sense if I write the polynomial as:

$$x^3 + 2x^2 + 3x + 5 = 0.$$

Then I say that this is a third-order polynomial. Very well; now I just need to learn how to solve it. As Papa always tells me, "Do not give up, *ma petite élève*." Even if I don't have a teacher like Antoine-August, I must try.

Sunday — June 28, 1789

Another protest shook the city with increased violence. Early this afternoon, while reading in Papa's library, I was startled by shooting and chanting. The angry voices jumbled with the sound of firearms shattered the quietness of the day. Another riot, I thought, and ran to the window. From there I spotted a mob of armed workers and market women, brandishing pikes, and walking westward. The demonstrators shouted slogans full of hate: "Hang the aristocrats!" and "Kill the despot rich; kill them, kill them!"

Angélique was terribly frightened. By the pallor of her face, I could tell Maman was also anxious. She shooed me away from the window and sent us all to her study. We waited until the mayhem subsided, wondering if Papa was caught in the middle of the riot on his way home.

The protests are occurring more and more frequently, and each time they turn more violent. Many people are angry with the king, seeing how he lives in luxury, seemingly oblivious to the suffering of his subjects. Workers are losing their jobs, and most of them resent the unfair taxation, blaming Louis XVI for not doing enough to improve their economic situation.

I hope the meeting of the deputies and King Louis in Versailles mitigates the tensions. The newspapers are full of articles written by liberals who promote the formation of a French republic. The liberal reformers want to take away the power of the king and form a government elected by the people. When Papa and his friends discuss these issues my mother just listens and does not say much, but she supports the king and expresses those feelings in private. She does not like the idea of taking away the king's power, as Papa suggests.

Mother is very fond of Louis XVI and Marie-Antoinette. She still talks about the royal wedding as if it happened yesterday, but it was in 1769. Mother tells us that Versailles was ablaze with fireworks and lavish feasts. The people of France were happy, celebrating the nuptials of the young prince Louis and Marie-Antoinette, who was fourteen years old at the time. Well, now they're both adults and, having become king and queen of France, they must have learned how to rule our nation.

Thursday — July 2, 1789

We were almost trampled by a mob! When we returned home from visiting Madame de Maillard, we found ourselves in the middle of a march on rue Saint-Honoré. The angry protesters were on their way to the Palais-Royal. That's the most popular gathering place where leaders of the revolutionary movement speak against the monarchy. The protesters were chanting songs of rage against the king.

We were trapped in the middle. The faces of the men and women in the crowd were contorted with anger. My mother tried to stay calm, but I noticed how nervous she was by the way she pressed her hands to her heart. We had to wait on the side of the street for the mob to continue their march before we could pass. Angélique kept innocently asking, "Maman, why are those people shouting? Why are they so angry?" but Maman just embraced her protectively. I was scared, too, at the sight of such rage and held my mother's arm.

When we returned, I escaped into my father's library, the only place I feel secure. The demonstration persisted into the evening. The windows of the study were open because of the summer heat, so I could hear the protesters passing by, inviting people to join them. They were chanting, "*Liberté, Égalité, Fraternité!*" Some were also intoning slogans of hate I wish I did not hear.

Papa and his colleagues had expected these demonstrations to break out at any time, as many workers are affected by France's failing economy. The causes of the people's turmoil and rebellion include the shortages of bread, the increase in food prices, and the loss of wages. Some leaders are threatening violence if the government does not do something to ease the workers' suffering. But nobody offers any practical solutions.

Yet, there's nothing I can do. And so I continue solving polynomial equations. The linear polynomial equation $ax + b = 0$ is the easiest to solve since $x = -b/a$. For example, $2x + 3 = 0$ has the solution $x = -3/2$. These are rather trivial; I would like to solve more challenging equations.

First, I need to learn the rules of algebra. But this will have to wait until tomorrow because it is already late. If Mother sees the light of my candle at this hour, she will be furious.

Wednesday — July 8, 1789

Who invented the symbolic notation of algebra? Who was the first mathematician to use a letter of the alphabet to represent the unknown quantity in an equation? In ancient times equations were written using word statements, since algebraic notation was not invented until much later. In the beginning, an equation was stated in words like this example: *A heap, its whole, its seventh, it makes 19.* A "heap" represented the unknown. With modern notation I could write this statement as an equation, using the letter x as the unknown: $x + \frac{1}{7}x = 19$. In this form, any mathematician in the world can interpret it, regardless of what language she speaks, and she would understand that the equation requires that one find the value for x that makes the equation true.

Diophantus of Alexandria was perhaps the first mathematician that made such problems more concise, leading the way for later mathematicians to develop the algebraic notation used today. I prefer to work with symbolic equations, but it is also a good exercise to start with a verbal statement and then rewrite it in symbolic or algebraic form.

The other day, just for fun, I asked Angélique, "Do you know when Papa will be twice as old as I?" My little sister did not know, of course, but I showed her that I could use algebra to answer the question. I can write an equation that includes what I know, and what I don't know, which I call x (the variable). Starting with the fact that Papa is 63 years old and I am 13, after x number of years Papa will be $(63 + x)$ years old, and I will be $(13 + x)$ years old.

I derive a mathematical expression to help me find out when Papa will be twice my age:

$$(63 + x) = 2(13 + x)$$

and then I solve for x and get

$$x = 63 - 2(13) = 37.$$

This means that in 37 years, Papa will be twice my age; I will be 50 years old $(13 + 37)$, and he will be 100 $(63 + 37)$. "And you, my dear sister," I said, "will be 47 years old!"

Très bien. These are trivial problems with very easy solutions. Now, suppose that I have to solve this equation: $2x^2 + 4 = 20$. In this case, the variable x is raised to the second power, so I have to use a square root to solve it. First, I divide each term by 2 to get $x^2 + 2 = 10$, and then I solve, $x^2 = 8$. Thus, after taking the square root, I get $x = 2.82\ldots$.

But wait—shouldn't I get two answers? Since a number x multiplied by itself yields the square of x, (or x^2), x can be negative or positive. For example, $x = 2$ results in $x^2 = 4$, and $x = -2$ also results in $x^2 = 4$. So, either negative or positive values of x will give the same square. This means that a second-degree linear polynomial equation will have two solutions. Should I assume the degree of the equation determines the number of solutions it has?

Tuesday — July 14, 1789

What a frightful day! A huge mob of angry workers stormed the Bastille prison and hundreds of people died. Papa said it was one of the most brutal and tragic demonstrations yet!

It all began this morning. My sisters and I were in Mother's study taking our daily lesson. We heard shooting in the distance and at first Maman paid no attention. But soon we heard the commotion getting closer and the racket in the streets became so unnerving that it was impossible to ignore. We ran to the windows and watched hundreds of armed people running, chanting and waving their fists. The multitude moved east towards the prison, chanting so stridently, "*A la Bastille! A la Bastille!*"

The siege of the Bastille, July 14, 1789.

After the mob disappeared from sight, we resumed the lesson. I didn't imagine the massacre that was about to happen. But when Papa came back from Versailles in the evening, he told us. A gruesome fight took place when the enraged workers arrived at the state prison. It was supposed to be just a protest against the increase of taxes. But the workers were defiant and the violent rally ended in a hideous massacre. The prison guards could not control the crowd, and many were killed. The Marquis de Launay, the governor of the Bastille, had his throat cut on the steps of the Hôtel de Ville, the town hall, and his head was paraded around the streets of Paris. I was shocked to learn of such horror.

Papa doesn't approve of violence, but he understands why the workers have taken to the streets of Paris demanding social rights. Father sees himself as a moderate revolutionary who owes greater allegiance to his nation than to his monarch, as he supports the ideals for social equality. My mother is unswerving in her loyalty to King Louis; she doesn't understand why he is blamed for what is happening. Of course she is sympathetic to the plight of the poor, but she doesn't believe that fighting is the way to solve the social problems in France. I agree.

Saturday — *July 18, 1789*

A new era for France begins. The Constituent National Assembly began to rule and immediately its members began drafting a new constitution. The king came from Versailles with some of his retinue, but without the queen. Louis XVI arrived to the Hôtel de Ville, where the new mayor of Paris, Monsieur Jean Sylvain Bailly, received him, surrounded by a large crowd of common citizens.

Mother wanted to go to see the king, but my father did not think it prudent to be near the city hall. He was afraid another violent demonstration could erupt since many people are unhappy with His Majesty. The meeting, however, went on peacefully. At one point, Papa related, the king placed the blue and red ribbon on his hat as a gesture of support for his subjects.

Monday — *July 20, 1789*

I study *Elements of Algebra*, a book written originally by a mathematician named Leonhard Euler. The book opens with a discussion of the nature of numbers and explains the signs + *plus* and − *minus*. Then it follows with the nature of whole numbers, or integers, with respect to their factors. The author states that algebra considers only numbers, which represent quantities, without regarding the different kinds of quantity.

When Millie came this evening, she seemed curious about the numbers on my paper. I was glad to talk with someone about my study of algebra. To make it simpler, I told Millie that when I use a mathematical statement to describe a relationship, I use letters to represent the quantity that varies, since it is not a fixed amount. These letters and symbols are referred to as variables. For example, it can describe the relationship between the supply of bread and its price.

It was hard to explain algebra to someone who doesn't read or write. I used other simple examples that Millie could understand, explaining that ancient mathematicians did not write equations like the ones she sees in my notes. The notation I use was not invented centuries later.

For example, if she'd ask how old I am, I could reply, "My age, plus 16, is equal to three times my age, minus 10." Using algebra, she could figure out my age by writing my words as an algebraic equation, using the letter x as my age, which is the unknown number:

$$x + 16 = 3 \cdot x - 10.$$

The first expression stands for "my age in years plus 16," written as $(x+16)$. This is equal to the second expression for "three times my age, minus 10," written as $(3 \cdot x - 10)$. The equation has a variable x, the unknown in this case, so the solution to the equation is the number that makes the equation true. To solve the equation I put all the terms that contain x on one side, and all the other terms on the opposite side:

$$x - 3 \cdot x = -10 - 16$$
$$2 \cdot x = 26.$$

The answer is: $x = 26 \div 2$, or $x = 13$ (indeed, this is my age). This means that replacing each occurrence of x with 13 makes the original equation true, since

$$x + 16 = 3 \cdot x - 10$$
$$13 + 16 = 3 \cdot 13 - 10$$
$$29 = 39 - 10$$
$$29 = 29.$$

Voilà! This verifies the solution!

I can make up any algebraic equation related to some unknown quantity. They are fun and easy. Millie was amused but did not want to see more examples. She had chores to do before going to bed. I was left alone to solve more equations. Of course, I didn't tell her that a mathematical statement is not just for variations of real objects. For example, an algebraic equation such as $x^2 + 1 = 0$ does not have to refer to anything in particular. That is why learning algebra requires abstract thinking.

Wednesday — July 22, 1789

I've been reading about the ancient mathematician Diophantus of Alexandria. Nobody seems to know exactly when he was born or when he died. The encyclopedia simply indicates that Diophantus lived about 250 A.D. Then I consulted the *Greek Anthology*, which is a collection of epigrams, and found among its mathematical problems a riddle about the life of Diophantus:

> God granted him to be a boy for the sixth part of his life, and adding a twelfth part to this, He clothed his cheeks with down; He lit him the light of wedlock after a seventh part, and five years after his marriage He granted him a son. Alas! late-born wretched child; after attaining the measure of half his Father's life, chill Fate took him. After consoling his grief by this science of numbers for four years he ended his life.

What I infer from this statement is that the son of Diophantus was born after $1/6 + 1/12 + 1/7$ of his life plus 5 years. The son died 4 years before Diophantus, and lived half as long. To solve the puzzle mathematically, I write everything in terms of the age of Diophantus, or the number of years he lived, and, since it is unknown, I call it x. Now, according to the riddle, his son was born in the year equal to $(1/6 + 1/12 + 1/7)$ times x plus 5. This can be written as

$$\left(\frac{1}{6} + \frac{1}{12} + \frac{1}{7}\right) x + 5.$$

I also know that the son died 4 years before Diophantus died, so I use the relation $(x - 4)$. I then subtract the year of birth from the year of death, and get the son's lifespan, which is half of his father's:

$$(x - 4) - \left[\left(\frac{1}{6} + \frac{1}{12} + \frac{1}{7}\right) x + 5\right] = \frac{x}{2},$$

This equation reduces to $\frac{9}{84}x = 9$.

Solving for x the age of Diophantus, I find: $x = 84$.

From this I deduce Diophantus married at age 33, had a son when he was 37, and died when he was 84. *Voilà!*

Friday — July 24, 1789

Monsieur du Chevalier is moving to Sweden! He did not come to give me a piano lesson today. Instead, he sent his valet to notify my parents that M. du Chevalier will no longer teach me because he is leaving France. Papa was not surprised. Members of the First Estate are angry with King Louis for taking away their privileges. Others are afraid of the revolutionary movement, and to avoid persecution, many aristocrats are emigrating.

Since my lesson was cancelled, Maman gave me permission to go to the bookshop down the street and buy a book. I love to go there; the smell of old books reminds me of the times I spent with Papa in his library when I was a little girl. He used to show me his childhood books and taught me arithmetic from them.

I was nervous to go alone to the bookshop because Monsieur Baillargeon, the proprietor, intimidates me. He is an old gaunt man with pale blue eyes and a deep frown on his forehead that makes him look perpetually angry. But I mustered the courage to ask for a book on mathematics. M. Baillargeon looked at me strangely, as if he did not understand me, and then whispered, "Perhaps mademoiselle wishes to buy a nice romance novel." I thanked him politely and repeated that I wanted a book *with* mathematics. After a few moments' hesitation, M. Baillargeon smiled condescendingly, but scurried to the back of the shop. He returned with an old dusty book and handed it to me.

I eagerly flipped through the yellowed pages. The book was old and fragile, but none of the pages seemed to be missing. With a few illustrations and many equations, the book looked interesting and cost only twenty-three *sous*. M. Baillargeon kept looking at me inquisitively, so I paid him and left in a hurry to get away from his prying gaze.

The book is a treasure. It's a translation of the *Arithmetica*, a famous work written by Diophantus hundreds of years ago. Diophantus, from the Greek city Alexandria, was perhaps the first to develop algebra. Ever since I read the story of Hypatia, I was curious about him. Hypatia wrote about *Arithmetica* and taught it to her pupils. That is why I must study it.

We are going for summer vacation tomorrow morning. My mother is anxious to leave the foul city air and refresh our spirits in the country. We're going to our house in Lisieux, a far distance away west from Paris. We normally depart in late June; this time, however, because of my father's business and his work with the National Assembly, we are delayed. I look forward to our holiday in the countryside. The voyage is long and tiring, but our home in Lisieux is a welcome respite from the summer heat in Paris.

Sunday — July 26, 1789
Lisieux, France

After an exhausting trip, we arrived in Lisieux. We left Paris yesterday morning and the long journey took a day and a half. At times, it felt hot and crowded in the rickety stagecoach, but I immersed myself in a book and tried to ignore my sister's incessant chatter.

We stopped at a tiny village to eat lunch and exchange the horses. After a short rest we resumed the trip and rode for another six hours. We arrived in Rouen at sunset. The inn where we spent the night was an old château, dark and ugly. Before dinner, Angélique and I visited the ancient chapel nearby and explored its decaying cemetery. While reading the inscriptions in the tombs, trying to imagine the lives of the people who were buried there, we were startled by an odd looking man. Toothless and with an ugly scar on his cheek, the old man resembled a character in a horror tale. Limping, the man approached us smiling, as if happy to see us.

After chatting nonsense, the old man pointed at a fairly new grave. He said the daughter of the Marquis who owns the inn was buried there. He added that the young girl died of fright one night when she met the ghost of the castle. He was probably making up the story, for he kept smiling mischievously. But Angélique was so terrified afterwards that she had to sleep in Maman's bed. In the morning, after an unappetizing breakfast, we continued our trip.

Lisieux is a quaint town in the *Pays d'Auge* region in northwest France. Our house is located outside the village, hidden among gentle hills, valleys, and forests. The landscape is beautiful, dotted with old manor homes, some châteaux, and half-timbered farm houses. The region is enchanting, and the smell of cider mixes with the aroma of cheese. As we approached the *Pays d'Auge*, the monotony of the long ride from Paris was broken by the sight of thatched roofs, winding brooks, charming chapels, and peaceful hamlets as colorful as those seen in old paintings.

We arrived in time for a delicious supper. Margarite, the lady who keeps the house, and her daughters had the table ready with cold meats, green salad, and refreshing lemonade. We spent the afternoon reacquainting ourselves with the village; we also attended evening Mass.

Within minutes after sundown, thousands of glittering stars began to pierce the sky, in a horizon that slowly changed from orange gold to dark. This is a sight I cherish.

It is late already, and Maman and my sisters have gone to sleep. The house is quiet, and I sit here, listening to the sounds of the country, watching the flame of my candle being swayed by the gentle breeze. A fresh bed with a down feather mattress awaits me to restore my tired body. Margarite put sachets of flowers in my pillow to attract happy dreams, she claims. Oh yes, I must rest and dream of what I'll discover in my books next.

Tuesday — August 4, 1789
Lisieux, France

I am studying the *Arithmetica* of Diophantus. It deals with the solution of algebraic equations and the theory of numbers. *Arithmetica* is a collection of problems with numerical solutions of indeterminate and determinate equations (those with unique solutions). Although parts of the book were lost, probably even before Hypatia's time, the parts that remained contain the solutions of many problems concerning linear and quadratic equations, but it considers only positive rational solutions to these problems.

Diophantus studied three types of quadratic equations: $ax^2 + bx = c$, $ax^2 = bx + c$, and $ax^2 + c = bx$. I read that Diophantus did not know the number zero, and so he considered the coefficients a, b, and c to be positive in each of the equations.

I understand why Diophantus did not include negative numbers. If one uses numbers for counting physical objects, like apples or houses, then "minus 3 apples" or "−7 houses" makes no sense. However, if one thinks of numbers as ideas, as mathematical objects, then one can readily accept numbers having negative values. For example, if I imagine numbers as points along an infinite line, then I must include negative numbers. With zero as the middle point on the line, I can see numbers along the right side of the line (positives) and numbers on the left side (negatives). Otherwise the line would end at zero, which makes no sense, unless zero was positioned at infinity!

But, if the solution to the simple linear equation $ax + b = 0$ was known since ancient times to be $x = -b/a$, then why didn't Diophantus include the negative numbers? The easiest way to interpret this equation is by writing $a = 1$ so that $x + b = 0$; the value of the variable x is simply the negative value of b ($x = -b$). For example, in the equation $x + 5 = 0$, x has to be equal to -5 in order to make the right hand of the equation zero. There it is, unmistakably stated; we need negative numbers to solve this equation.

Friday — August 14, 1789
Lisieux, France

Solving certain polynomials is rather easy. I know that polynomial equations consist of terms involving an unknown number x raised to a positive whole power or exponent and multiplied by constants a and b. Such equations are classified by the largest exponent of the unknown x.

First-degree equations are those like $ax + b = 0$. Second-degree equations are like $ax^2 + bx + c = 0$. Third-degree equations are like $ax^3 + bx^2 + cx + d = 0$, and so on. Either of the constants can be equal to zero, in which case the number of terms is reduced. The important thing to look for is the largest exponent of the unknown number.

Solutions to this kind of polynomial equations will always be algebraic numbers. Algebraic numbers are the roots of polynomial equations with integer coefficients. For example, linear equations that have the general form $ax - b = 0$, where a and b are two known numbers, have solution $x = b/a$. This would be a fraction, zero, or an integer, depending on the values of a and b. For example, $x = 2/3$ is algebraic since it is the root of $3x - 2 = 0$. On the other hand, $\sqrt{2}$ is also an algebraic number since it is the solution to the equation $x^2 - 2 = 0$.

I've solved many first-degree equations. They're very easy because the equations have a single variable, say x, raised to the first power. For example, $ax + b = 0$, in which x is the same as x^1. In other words, x means "x raised to the first power."

I also solved equations of the form $ax^2 - b = 0$. The solutions are of the form $x^2 = b/a$. Again, depending on the values of a and b, I get a solution that is an integer, a fraction, or a radical. I solved many equations, every one of which I could find in my books, and then made up some myself.

Now, let's consider a circle of diameter d and perimeter p. If I divide the perimeter by its diameter, the result always gives the same number, and this number is π, no matter how big or small the circle is. In other words, $p/d = \pi$. I know that π is *not* a rational number. Does it mean that π is not an algebraic number?

Thursday — August 27, 1789

Our summer vacation ended abruptly. We returned to Paris yesterday after the riots started in the provinces. My mother became very nervous when the signs of unrest in the village threatened the peace around us. The revolutionary movement is spreading very quickly, causing rioting and pillaging in places that used to be so quiet. Incited by a few, the farmers who worked for the rich landowners all their lives are taking back the land, using all means possible, including pillaging and violence. We witnessed a family fleeing their estate, terrified by the angry mob of peasants raiding châteaux and burning land titles.

Paris is also in a state of chaos. Last night, the archbishop's house was vandalized. We heard windows breaking and the mob's shouting woke us up around three in the morning. Millie saw later in the daylight that the property was badly damaged. The workers were furious with the rich aristocrat abbots who live in luxury while they live in misery. The people resorted to violence to express their frustration.

Papa brought home a copy of *The Declaration of the Rights of Man and of the Citizen* just issued by the National Assembly. The declaration states that all men are born and remain free and equal in rights. It also implies that all citizens of France will share the political power. However, it only mentions men. Why are women not included? Don't women have the same rights? I should have asked Papa, but I didn't have a chance. My sister Madeleine came in with her fiancée Monsieur Lherbette and everybody started to congratulate them on their engagement. They will marry sometime this year.

Antoine-August came to visit with his parents. While the adults conversed, he showed me several equations he has to solve for an assignment, such as this: $3x^2 - 4x = 2$. This and the other equations are of the general form known as the quadratic equation $ax^2 + bx + c = 0$, where a is a nonzero number and x is the unknown variable. I said that one way to solve them is by applying the quadratic formula I've learned:

$$x = \frac{-b \pm \sqrt{b^2 - 4ac}}{2a}.$$

So, to show him how to use it, the first thing I did was rewrite the first equation in the form of the quadratic equation

$$3x^2 - 4x - 2 = 0$$

where $a = 3$, $b = -4$, $c = -2$. Then I showed him this substitution to get the solution:

$$x = \frac{-b \pm \sqrt{b^2 - 4ac}}{2a} = \frac{-(-4) \pm \sqrt{(-4)^2 - 4(3)(-2)}}{2(3)} = \frac{4 \pm \sqrt{40}}{6}.$$

I should say the *two* solutions because x has two different values, $x = \frac{4+\sqrt{40}}{6}$ and $x = \frac{4-\sqrt{40}}{6}$. Either value of x solves the equation.

These quadratic equations are very easy; I am sure Antoine-August could have solved all his quadratic equations without my help, but I am glad he asked.

Saturday — September 5, 1789

Mathematics is fascinating. It is the science of numbers and equations that requires the application of rules and logical thinking. I learn more mathematics every day by solving more problems. And I discovered that there are different ways to solve them. Just by observation I know that $x^2 - 9 = 0$ has two roots, $x = \pm 3$. I must prove it, of course. Other equations can be solved by factoring. For example, $x^2 - x - 2 = 0$ is factored as

$$x^2 - x - 2 = (x + 1)(x - 2) = 0.$$

Then I equate each binomial expression to zero and solve them individually:

$$x + 1 = 0; \quad x = -1$$
$$x - 2 = 0; \quad x = 2.$$

Of course, I could solve this equation using the quadratic formula, but it is much faster and easier to do it by factorization.

Moreover, if a second order equation can be written as a perfect square equal to a constant, then it could be solved easily. By perfect square I mean that the equation is factored as $(x + a)^2$. For example, $(x + 3)^2 = 5$ can be solved as follows:

$$(x + 3)^2 = 5$$
$$x + 3 = \pm\sqrt{5}$$
$$x = -3 \pm \sqrt{5}$$

so the roots (or solutions) are $-3 + \sqrt{5}$ and $-3 - \sqrt{5}$.

By the way, $(x+3)^2 = 5$ can be written in the quadratic form, $x^2 + 6x + 4 = 0$.

Now let's say I need to solve $x^2 + 8x = 1$. First I seek a constant k such that the addition of k^2 to both sides of my original equation yields a perfect square equation. I write

$$x^2 + 8x + k^2 = (x+k)^2 = x^2 + 2kx + k^2.$$

Then I must have $8x = 2kx$, from which $k = 8/2$ and $k^2 = 16$.

Now I add k^2 to both sides of my original equation $x^2 + 8x + k^2 = 1 + k^2$, or

$$x^2 + 8x + 16 = 1 + 16,$$

and I rewrite it as a perfect square equal to a constant and then solve:

$$(x+4)^2 = 17$$
$$x + 4 = \pm\sqrt{17}$$
$$x = -4 \pm \sqrt{17}$$

So, the two roots or solutions of the equation are $-4 + \sqrt{17}$ and $-4 - \sqrt{17}$.

Antoine-August asked which method is better to solve quadratic equations and I said that it depends on the equation. All the methods are correct, but a mathematician decides which one to use based on preference, or whether one wishes to do it in the simplest or fastest way possible.

He also mentioned cubic polynomial equations assuming that I am familiar with them. I replied, rather arrogantly, that if I could solve quadratics then I should find solutions for cubics as well. That was an impulsive response as I don't know, and so I must hurry to learn cubic equations. Otherwise Antoine-August would never ask me for help again.

Friday — September 11, 1789

Cubic equations are more difficult. Starting with the polynomial $a_n x^n + \cdots + a_3 x^3 + a_2 x^2 + a_1 x + a_0$, if the exponent is $n = 3$, the equation becomes $ax^3 + bx^2 + cx + d = 0$ and is called "cubic." These equations were fully solved hundreds of years ago using geometrical methods. However, solving the cubic equation using pure algebra was a great challenge to mathematicians. In fact, it proved so tough that the Italian mathematician Luca Pacioli wrote in his book *Summa de arithmetica* that "the general cubic equation is unsolvable."

The solution was obtained in the sixteenth century. Before that, mathematicians could solve only special cases of the cubic equation. For example, Scipione del Ferro could solve the depressed cubic, $ax^3 + cx + d = 0$. He kept it a secret because, at that time in Italy, new discoveries were used as weapons against opponents in mathematical contests. Just before his death, del Ferro handed the solution to his student, Antonio Fior. Then Niccolò Fontana, better known as Tartaglia, challenged Fior. To respond to Tartaglia's challenge, Fior retaliated with thirty problems on depressed cubics. Eventually, after much effort, Tartaglia solved the problems.

It is inspiring to know that a mathematician should fight so passionately for the honor of being the first to solve a difficult problem. It makes me more determined to learn how to solve these equations and try to be the best. Interestingly, most people knew Niccolò Fontana by his nickname Tartaglia, which means in Italian "one who stutters," because of his speech impairment. I read somewhere that he stuttered due to a wound he suffered in a duel.

Now I can also answer the question, is π an algebraic number? I can categorically say no. I reason as follows: by definition π is equal to the ratio of the circumference of a circle to its diameter, no matter how one measures it. If π were an algebraic number, then it would be a solution to some polynomial that has integer coefficients. It is not. Thus, if π is not an algebraic number, then it must be a number that goes beyond, that transcends the algebraic numbers. I know that π is not a whole number or a perfect fraction, and that the decimal part does not show a repeating pattern.

Oh, I am digressing. I must learn how to solve cubic equations before Antoine asks for help again.

Friday — September 18, 1789

Still struggling with cubics! At first I thought cubic equations were easy to solve. But soon it became clear they are more challenging. At least the procedure is not as straightforward as that for solving quadratic equations! About 200 years ago, an Italian mathematician named Girolamo Cardano published the solution to the cubic equation. I wish I knew it because I've tried solving a simple cubic and failed. It is so frustrating!

Cubic equations have the form $x^3 + ax^2 + bx^1 + cx^0 = 0$. Two nights ago, I attempted to solve a simpler cubic like this: $x^3 + mx = n$. It looked easy enough but, after trying for several hours, I couldn't find a solution.

What am I missing? I will not sleep well if I do not solve it. I will attempt a different approach.

First I see that $(a - b)^3 + 3ab(a - b) = a^3 - b^3$. Then, if a and b satisfy $3ab = m$, and $a^3 - b^3 = n$, then $a - b$ is a solution of $x^3 + mx = n$. Now $b = m/3a$, so $a^3 - m^3/27a^3 = n$, which I can also write as

$$a^6 - na^3 - m^3/27 = 0.$$

This is like a quadratic equation in a^3,

$$(a^3)^2 - n(a^3) - (m^3)/27 = 0.$$

So, I can solve for a^3 using the formula for a quadratic equation.

Très bien. Then I can find a by taking cube roots. I also can find b in the same way, or by using $b = m/3a$. Therefore, $x = a - b$ is the solution to the cubic equation.

Am I correct? Well, I substitute x into the original cubic equation and see if it's true. I could ask Antoine-August to show my analysis to his tutor. He would tell me if I'm doing this correctly.

Friday — September 25, 1789

Autumn is slowly making an entrance. The weather is so nice, and the colors of the foliage are changing. After lunch my sisters and I went for a stroll to the *Jardin des Tuileries*. As we made our way, Madeleine's fiancé, Monsieur Lherbette, came to join us. His company made the outing more enjoyable. The streets are swarming with indigent people, barefoot children, and haggard women begging for money. I noticed some of them angrily shaking their fists as the wealthy carriages passed them by. But others seemed too weary to notice. The sight of poor children is heartbreaking. I gave the few coins in my purse to a pretty little girl dressed in rags. The poor child, she didn't say a word but eagerly took the coins and ran.

When we were near the Louvre Palace, M. Lherbette told me that the French academies have residence there, including the Royal Academy of Sciences. There is where the best mathematicians in the world work on their research. I was thrilled to know I was walking so near the great men of science. M. Lherbette added that every week the learned society holds public lectures where the scholars expound their work. Wouldn't it be great if one day I could attend the lectures? As we went past the massive front doors of the old edifice, I saw gentlemen with stern faces coming and going, but I could not discern who they were.

One day, I know, I will meet a great mathematician, and perhaps he'll invite me to attend the lectures.

Thursday — October 1, 1789

To be a mathematician one must prove theorems. There are different ways to prove a mathematical statement. For example, the ancient Greek scholars developed a mathematical method to prove theorems that they called *reductio ad absurdum*, which means "reduction to absurdity." With this procedure, one assumes the opposite of what one desires to prove. Then one can show that this assumption leads to a contradiction. Or, if the opposite of what one wishes to prove is false, the statement to be proved must be true.

To learn how it works, I'll use *reductio ad absurdum* to prove the irrationality of numbers. I could start by assuming that a number is represented by some ratio of whole numbers, say p/q. If I can find a fraction p/q equal to the number, then the number is rational; however, if p/q is not equal to the irrational number, then I can conclude that no fraction exists equal to the number, and therefore it is an irrational number. It seems simple enough; I am sure I can prove that the square root of 2 is an irrational number.

For now I will try a game of numbers. I start with perfect squares. By definition, a perfect square is a number that can be written as a whole number times itself. For example, 81 is 9×9, and 25 is 5×5. Thus, 1, 4, 9, 16, 25 are perfect squares. The Greeks proved that any square root of a whole number that is not a perfect square is incommensurable. Therefore, the square roots of 3, 5, 6, 7, and 8 are incommensurable numbers. I also find that the sum of the first n odd numbers is a perfect square, that is $n^2 = 1 + 3 + 5 + 7 + \cdots + (2n - 1)$. I notice that every perfect square ends in a 0, 1, 4, 5, 6, or 9.

Oh, it must be very late; the clock in the drawing room chimed the eleventh hour long ago. I must sleep or tomorrow I won't get up on time for morning prayers. Maman would be terribly upset with me.

Wednesday — October 7, 1789

The court has moved to Paris. It was not a voluntary move on the part of the king—he was *forced* by the people. It began with the assault of the royal

Arrival of the royal family in Paris after the assault in Versailles, October 6, 1789.

palace in Versailles two days ago. A mob of angry working-class women marched to Versailles demanding to be heard by King Louis. They went to complain that the poor people in Paris can't afford to buy bread.

The demonstrators wanted to talk with His Majesty, to let him know they need help to resolve the economic problems. Armed with pitchforks, the women walked for six hours to Versailles. Upon arrival, the enraged women stormed the château, ran up the queen's staircase, and broke into the royal chambers. Queen Marie-Antoinette had to run to join the king and their children. How terribly scared she must have been, seeing hundreds of angry people brandishing firearms and pikes, shouting insults and running through the palace.

We know all this because Papa witnessed the assault. He said that there were mothers driven to despair because they could not feed their children. After storming the palace, a delegation of women gained an audience with the monarch; they also addressed the National Assembly. The protesters complained that the wealthy citizens are hoarding grain; some claim that is why there isn't enough bread for the poor.

My mother blames the leaders of the revolutionary movement for this revolt. She is a generous woman with a tender heart. In fact, she works with the *Charité Maternelle*, the charity founded by Mme Fougeret, to help poor women care for their babies. So, when she heard about the women marching to Versailles demanding bread, she understood their plight. However, she thinks someone incited the women to violence. We do not know how all this started.

King Louis promised to produce bread for everybody, and he allowed the women to escort him, his family, and his court back to Paris. This way, the king will be within reach of the people. The women assured His Majesty that in Paris he would find faithful advisors who can tell him how things stand with his subjects, enabling him to act accordingly. Louis XVI and his family now reside in the Tuileries Palace nearby. The National Assembly also moved to a building close to the Tuileries. I am so glad because Papa will not have to go to Versailles anymore. He'll be back home, and I will tell him about my studies.

Wednesday — October 14, 1789

To prove the irrationality of the square root of 2, I use *reductio ad absurdum*. I start the proof by contradiction, assuming that the square root of 2 is rational. Then I can express it as a ratio of two numbers p and q, which are the smallest positive integers that satisfy the relation

$$\sqrt{2} = \frac{p}{q}.$$

Of course p and q cannot both be even because, if they were, I could divide each by 2 and have a smaller p and q to work with.

From this I can get, by squaring both sides:

$$2 = \frac{p^2}{q^2}.$$

Then by multiplying by q^2

$$2q^2 = p^2.$$

From this I can deduce that p^2 is even.

If p^2 is even, then q must be odd since p and q cannot both be even. Since p^2 is even, then p must be even. If p is even, then p^2 must be divisible by 4. Thus, $\frac{p^2}{2}$ is an even integer. However, since $2q^2 = p^2$, I should write,

$$\frac{p^2}{2} = q^2.$$

Thus, q^2 must be even, and therefore it follows that q is even. But I had previously determined that q must be odd. And since q cannot be both even and odd, I conclude that no such integer exists, which invalidates my original premise that the square root of 2 is rational. In other words, $\sqrt{2}$ has to be irrational. That's it; I proved it!

Saturday — October 24, 1789

How did Archimedes confirm that the crown was not made of pure gold? According to the story, Archimedes discovered a method to prove it when he observed that the amount of water that overflowed the bathtub was proportional to the weight of his body submerged in the water. But how? I suppose Archimedes used algebra to solve the problem. If I had to do it myself, this is what I would do:

I start by assuming that I know the weight of the crown and the volume of water displaced by it, quantities that I call m and V respectively. If, as the king suspected, the crown was made of gold and silver, the crown weighed an amount equal to the sum of the weight of the two metals, say $m = m_1 + m_2$. This gives one equation and two unknowns. Thus, I also must know the amount of water displaced by a certain amount of gold and a certain amount of silver. I could say that the total volume of water displaced by the crown is equal to the contributions from the volumes displaced by each one of the two metals.

I need to know that the specific volume of a body is equal to its volume divided by its mass, $v = V/m$, so assume for gold $V_1 = v_1 m_1$ and for silver $V_2 = v_2 m_2$. Thus, I can write another equation relating the volumes of gold and silver making up the total volume of the crown, $V = V_1 + V_2 = v_1 m_1 + v_2 m_2$. Therefore:

$$m = m_1 + m_2$$

and

$$V = v_1 m_1 + v_2 m_2.$$

Now I have a set of two equations and two unknowns, m_1 and m_2. The simplest way is to solve for m_1 from the first equation, and substitute this into the second as follows:

I rewrite the first equation as

$$m_1 = m - m_2$$

and I substitute it into the second

$$V = v_1 m_1 + v_2 m_2 = v_1(m - m_2) + v_2 m_2.$$

So,

$$V = v_1 m - v_1 m_2 + v_2 m_2 = v_1 m + m_2(v_2 - v_1)$$

therefore

$$m_2 = (v_1 m - V)/(v_2 - v_1)$$

And since m, V, v_2, and v_1 are known quantities, now I know the weight of silver, m_2. Substituting it into $m_1 = m - m_2$, I obtain the weight of gold. Obviously, if the crown were made of pure gold the value of m_2 would be zero.

Is this how Archimedes proved that the crown was made of gold and silver?

Tuesday — November 17, 1789

I have not written in a while because Mother confiscated my paper and pens. She was angry with me after she found me studying all night. I couldn't believe it myself, but the time goes by so fast when I am solving problems. Maman gave me a harsh lecture, again. This morning, Papa talked with her and convinced her that my studies are harmless, as long as I do not stay up for many hours. Reluctantly, Maman agreed to give me back my pens. But I had to promise her I would snuff my candle and go to sleep no later than eleven o'clock.

I continue my study of numbers. Numbers are the foundation upon which mathematics rests. And the nature of numbers is established by rigorous rules, because not all the numbers are the same. For a merchant, numbers are tools to perform the basic arithmetic computations like adding and subtracting. Most people rely on natural numbers such as the integers 1, 2, 3, ... that is, the ordinary numbers that one uses for counting, adding, and subtracting objects. Of course, the negative numbers and zero can also be considered ordinary numbers, but most people would not think much about negative numbers since they do not use them for everyday computations.

The rational numbers, which are ratios of integers such as $1/2$, $5/2$, $13/27$, are the numbers that the Pythagorean mathematicians believed to rule the universe. This type of number has special rules, such as the one that establishes that a rational number a/b is valid only if b is not equal to zero.

In other words, division by zero is never allowed. There are mathematical theorems for rational numbers that prescribe this rule.

One theorem states: "The product of any rational number c and zero is equal to zero, or, $0 \cdot c = 0$." Another theorem states: "$a/b = c$ if and only if $a = b \cdot c$." This theorem says that if the rational number a/b is equal to another number c, then it must be true that the numerator of a/b equals the denominator multiplied by c. For example: $12/3 = 4$ if and only if $12 = 3 \times 4$. Now, if I write $12/0$, this would imply that there is a number c equal to $12/0$, and by the second theorem I would have to write $12 = 0 \cdot c$. However, the first theorem tells me that the product of any rational number c and zero is equal to zero. How can I have $12 = 0$? This is absurd. Therefore, I conclude that no rational number can have zero as its denominator. This is how the ancient mathematicians established the rigorous rules by which mathematics is built.

I performed endless calculations, starting with ratios such as $5/2 = 2.5$ and $3/4 = 0.75$, but then I discovered that relatively simple rational numbers such as $5/3$ and $7/3$ have infinite repeating decimals, since $5/3 = 1.666666\ldots$. Then I came across other types of rational numbers such as $2/7 = 0.285714285714\ldots$, a number whose integers after the decimal point repeat over and over indefinitely.

It wasn't just my curiosity about how many digits I could calculate. I also wanted to understand the difference between rational and irrational numbers. There is a theorem that states: "any irrational number is represented by an infinite decimal where the string of digits neither terminates nor has a repeating pattern." I used this fact to conclude that the number $2/7$ is not irrational since its decimal has a repeating pattern. However, the square root of 2 and π are irrational numbers since $\sqrt{2} = 1.41421356\ldots$, and $\pi = 3.14159265\ldots$, and it appears that the digits of the decimal part are randomly distributed.

When I do mathematics I lose track of time; I transport myself to another world where nothing else matters. I must be careful, or I could stay up all night and would lose my mother's trust. I promised her to go to bed early.

Saturday — November 21, 1789

It is so cold already. Yesterday, a poor woman froze to death at the steps of the church. How ironic that the archbishop lives in a luxurious palace just a few meters away from where the wretched woman died of hunger and cold.

Couldn't he have done something for her? People should not die in that way at the entrance of a sacred place. It is sad and inhumane.

Papa believes that the economy and the social conditions of the poor will improve now that the National Assembly nationalized all church property. This was the Assembly's solution to the increasing economic problems France now faces. The delegates did not want to levy new taxes; instead, they proposed to take over the church lands. The Assembly plans to sell the lands and issue bank notes that will become the new French currency. It seems fair to me that the wealth of the Church is used to help the poor people.

Saturday — November 28, 1789

I'm in trouble! Mother is so angry at me that she took away my candles. A few days ago she found me awake, solving equations, unaware that it was past midnight. Maman was enraged that I disobeyed her, and as punishment she now sends me to bed in darkness. I am only allowed to read in the library during the day. That's fine. I'll use that time to learn more mathematics.

I am writing by the light of an old worn candle I found in the kitchen. It would be much easier if Maman were more understanding. I would not have to sneak out in the middle of the night looking for candles. And now that winter is approaching, my bedroom gets very cold at night. Oh Mother, would you ever accept my desire to learn mathematics? Taking away the heater in my room or the candles will not diminish it.

Saturday — December 5, 1789

Paris is quiet and peaceful tonight. The calm effect is soothing, as if the violent events of the previous months never happened. The cold of winter forces people to stay home, so the political instigators have smaller audiences. December is also the time of year when people are in a cheery mood, and families focus more on the Christmas festivities. Whatever the reason for this tranquility, I am glad that the fighting and hostilities we witnessed in the summer have vanished. Paris is again the beautiful city I love. Even the glow from the lamp posts seems gentler and warm.

I like this time of year because Paris becomes an enchanted city, lit by thousands of candles, and there is merriment everywhere. I hope the riots of the summer don't happen again.

Wednesday — December 9, 1789

Papa came to see me, worried because I was absent at the dinner table. Yes, I was upset because my mother gave me a lecture. Another one! She chastised me for spending too much time reading and doing all this *crazy* mathematics, as she calls my studies. Her words hurt me, and that's why I did not feel like eating. I also did not want to run into Madame Charbonneau, who came to visit my mother. She is a silly old woman who always finds reasons to criticize me, especially if she sees me reading, or playing chess. She remarks about my appearance, saying that I don't look feminine, just because I do not put rouge on my cheeks.

I am so glad Papa came to see me. His conversation cheered me up, as always. I showed him my notes, and he was astounded. My father kept looking at the equations I've written and couldn't believe I had done all that analysis by myself. He was astonished and proud of me when I showed him the solution to the riddle he gave me in April. I solved it weeks ago, trying to master algebra. Here I rewrite the riddle followed by my solution:

A long time ago, in a far away land, a sage woman raised three boys to be honest and smart. When her husband died, he bequeathed her one milking cow, and thirty-five more to be divided among their three sons. According to the wishes of the father, the oldest was to receive half the number of cows, the second would get a third of the total, and the youngest would receive a ninth of all.

The brothers began to quarrel amongst themselves, as they did not know how to divide the herd. They attempted different ways to divide it, and the more they tried, the angrier they became. Every time one of the brothers proposed a way to divide the animals, the other two would shout that it was unfair.

None of the suggested divisions was acceptable to all. When the mother saw them squabbling, she asked why. The oldest son replied: "Half of thirty-five is seventeen and one-half, yet if the third and the ninth of thirty-five are not exact quantities either, how could we divide the thirty-five cows amongst the three?" The mother smiled, saying the division was simple. She promised to give each their share of the herd and assured her sons it would be fair.

The woman started by adding her own cow to the herd of 35. The sons were stunned and thought she was foolish to give away her only cow. But their mother assured them there was no need to worry, as she would benefit the most.

The woman began: "As you see, my sons, now there are 36 cows in the herd." Turning to the oldest, she stated that he'd get one-half of the 36, which is 18 cows. "Now you cannot argue, as originally you would have gotten 17 and one-half. You gain with my method, right?"

Turning to the second, the mother continued: "You, my middle son, will get a third of 36, which is 12 cows. You would have gotten a third of 35 which is 11 and a little more of another, right? You cannot argue either, as you also gain."

And to the youngest son the woman said: "According to your father's wishes, you were to get a ninth of 35, which is 3 cows and a fraction of another. However, I will give you a ninth of 36, which is 4 cows. You also gain and should be satisfied with your share."

The wise woman ended with these words: "With this division that favors all, 18 cows are for the oldest, 12 for the middle, and 4 for the youngest son. This gives a total of 34 cows $(18 + 12 + 4)$. Of the total 36 cows, 2 remain. One, as you know, belongs to me anyway; the other, well, I think it fair that I also keep, as I have divided your inheritance in a manner that is fair for all.

Papa's eyes shone brighter when he said, that's exactly the answer to the mathematical riddle! He was pleased and very proud of my accomplishments. However, I have a great deal more to learn. If it took centuries for mathematics to develop, I can just imagine how many more concepts there are that I do not yet know.

Tuesday — December 15, 1789

The foundation of algebra was established in ancient Egypt and Babylon hundreds of years ago. Of course, the ancient mathematicians did not write equations as we do today. For example, Euclid developed a geometrical approach that involved finding a length, which represents the root x of a quadratic equation. He had no notion of equation or coefficients; Euclid worked with purely geometrical quantities. Also, the Babylonians developed an algorithmic approach to solving problems, which could be interpreted as the quadratic equation we know now. Their solutions were always positive quantities, since they represented lengths and areas, just like Euclid's method. That is why negative numbers were not known in antiquity.

The word "algebra" is derived from the Arabic word for "restoration," *al-jabru.* Historians find that Arab mathematicians knew the solutions of equations as the "science of restoration and balancing." Ancient mathematicians wrote out algebraic expressions using only occasional abbreviations, but by medieval times mathematicians were talking about arbitrarily high powers of the unknown x, and they worked out the basic algebra of polynomials, without using modern symbolism, of course. A Latin translation of Al-Khwarizmi's *Algebra* appeared in the 1100s. A century later, the Italian mathematician Leonardo Pisano, better known as Fibonacci, was able to find solution of some cubic equations.

Early in the sixteenth century, the Italian mathematicians Scipione del Ferro, Niccolò Tartaglia, and Girolamo Cardano solved the general cubic equation in terms of the constants appearing in the equation. Ludovico Ferrari, who was a pupil of Cardano, soon found an exact solution to equations of the fourth degree, and since then mathematicians have tried to find a formula for the roots of equations of degree five, or higher.

Also in the sixteenth century, algebra was further developed with the introduction of symbols for the unknown and for algebraic powers. The French philosopher and mathematician René Descartes introduced such symbols in his Book III of *La Géométrie.* Descartes' geometry treatise also contains theories of equations, including the rule of signs for counting the number of what Descartes called the "true" (positive) and "false" (negative) roots of an equation. In his book *Elements of Algebra*, Euler defined algebra as "the science which teaches how to determine unknown quantities by means of those that are known."

Many mathematicians developed algebra in a period of many centuries. Each one made a contribution to the current knowledge, like adding a gold coin to a treasure chest. Right now it looks plentiful, glistening in the sun. However, the chest of knowledge is not full. Mathematics, I am sure, still holds many mysteries that need to be unraveled and an infinite number of equations to be solved. That is why I wish to become a mathematician to contribute to the treasure chest of knowledge.

Friday — December 25, 1789

What a wonderful day this was! Starting after morning prayers, everybody was in a joyful mood, hugging and kissing, wishing one another good things. My sisters and I helped Maman set up *La Crèche*, the Christmas crib, with figurines representing the shepherds, the farm animals, and the holy family. The three wise men will be added to the nativity at Epiphany in January.

Early in the day, we had a special Christmas lunch. We started with fresh vegetables and *anchoiade*, a salty dressing made with anchovies that I dislike but ate nevertheless. It was followed by a delicious spinach omelette, fish with tomato sauce, and goat cheese. Then we ate bits of the thirteen Christmas desserts, fruits and walnuts, and Provençal wine. Angélique asked, as she does every year, why there are thirteen. So Maman explained *again* that the thirteen desserts represent Jesus and the twelve apostles.

And as she does every year, Maman placed three candles on the table to represent the Holy Trinity. I know this is ridiculous, but Madame Morrell tied the corner of the tablecloth with a knot so that, according to her, the devil can't climb on the table. After lunch Mme Morrell insisted on keeping the table set until after the Christmas Mass, so that "the spirit of the Saints can eat what is left." Although we do not believe in those silly superstitions, my parents do not object; like other harmless customs, they are part of the traditional festivities of people from the villages.

After lunch we visited friends and relatives. For dinner, Mme Morrell prepared mutton, lamb chops, asparagus soup, boiled vegetables, and cheese. Then we dashed to *La Pastorale*, the shepherd's play, in the courtyard of Notre-Dame. The performance finished in the church before midnight Mass. The ceremony was beautiful; there were hundreds of candles with flickering flames casting a golden glow on the statues of saints and virgins. The smell of incense was overwhelming. The music made me shiver and more so when the chorus sang.

The ritual ended past midnight. Upon leaving the church, we drank a sip of warm wine Papa bought from an old woman by the bridge. The night was cold when we returned home, but it did not bother me. During the ride home, a pale moon cast its shimmering light on the city. I looked up and discovered a dark sky full of blinking stars. How happy it made me feel. At home we gathered at the dinner table once more to express good wishes to each other.

Paris, France
1790

2
Discovery

Saturday — January 2, 1790

I sit in Papa's study, watching through the window the velvety white snow blanketing the city. The afternoon light intermingles with the shadow of the night. The street lamps are lit, adding a golden glow to the twilight. People go about their affairs, poor and rich, conscious of each other's social status. The wealthy travel in their fancy carriages, bundled up in fine furs, while the less fortunate trudge in the muddy streets, wearing worn-out garments to protect them from the cold wind. Rich and poor, young and old, all humans are alike in their desires, but they are not socially equal.

I never thought about it before, but now I see that the citizens of France do not share equal rights. Last year the workers and poor peasants revolted, unhappy with the unfair tax structure, and the financial crisis that led to an increase in prices. Bread became scarce and the poor went hungry. While the royal family and the court were living in luxury at their Palace in Versailles, the peasants and the workers in the city were suffering.

Papa and his friends criticize Louis XVI for being an incompetent ruler. The king is good, has a noble heart, but lacks leadership and is not in touch with his subjects, especially the lower classes. Papa asks, how is it possible that a feeble monarch who seeks only to satisfy his own pleasures can govern a nation? And many people dislike Marie-Antoinette because she was capricious and irresponsible in her early years as queen. Rumors about her abound, that she spends huge amounts on clothes while the poor people starve outside her palace. Other people loathe the queen simply because she is Austrian.

Now I also understand why the people forced the king and his court to move to Paris. They hope the king will be more effective helping his subjects. And he is trying. Louis XVI is attempting to solve the financial crisis by removing some of the exemptions on taxes. The aristocrats are furious. His Majesty faces strong opposition from the nobility. That is the reason he called the meeting of the Estates General last year. However, very little has changed for the poor people since then. Even after moving to the Tuileries Palace in Paris, the king cannot resolve the social conflicts and the financial crisis of the nation.

What will happen next? Will 1790 be a better year? I close my eyes and make a wish, as I used to do when I was a little girl. May the New Year bring us peace.

Wednesday — January 6, 1790

Winter is my favorite season, perhaps because the city becomes quiet, as if life stands still. The white snow covering the roofs makes the buildings look enchanted. In winter I spend more time in solitude and study. In the coldest days, I huddle in my father's library in front of the fireplace. This evening, to entertain my sister I made up the following story:

> There was once a king who owned many of the most gorgeous palaces ever built. He also had several daughters who one day would rule the immense kingdom. When he died, the monarch left the palaces to be divided among all his daughters. In his testament, the father stated that the division of property should be as follows: the oldest daughter would receive one palace and a seventh of the remaining buildings. The second daughter would get two palaces and a seventh of what remained. The third girl would receive three palaces plus a seventh of the rest. And so on until all his daughters received their share.
>
> The princesses were unhappy, believing some were getting less than the other sisters. They went to their mother crying. The queen, who was an intelligent and wise woman, responded that their protest was unfounded, as their father had given all his daughters a fair share of the kingdom.

Can you tell how many princesses the king had, and how many palaces each one inherited?"

Angélique guessed the answer, but incorrectly. This is how one can solve the riddle:

The first princess received 1 palace and a seventh of 35, which is 5. That is, she received 6 palaces, leaving 30 for her younger sisters.

Of the 30 palaces, the second princess got 2 plus a seventh of 28, which is 4. She was given 6 palaces, leaving 24 for the remaining sisters.

One can go on doing this analysis in the same manner until all palaces are given. So the correct answer for the riddle is: the king had 6 daughters and left them 36 palaces to be divided equally.

The problem can be solved with algebra, stating it with a simple formula that has two unknowns. Let the number of palaces be x, and the number of princesses $(n-1)$, and one can derive the following formula: $x = (n-1)^2$.

From there I determine that the first princess would receive 1 palace and $1/n$ of the rest. The second would get 2 palaces plus $1/n$ of the rest. And so forth. One can prove it by solving for the case in which $n = 7$, so that $x = 36$.

Monday — January 11, 1790

It is, at this very moment, bitterly cold. My fingers are so numb that I cannot feel them, and my chest feels tight with this overwhelming chill in my room. All around me is quiet. No screech of steel-tired wagon wheels is heard in this wintry night. The city is dark and quiet. The ink in my pen begins to thicken. I must put away writing to get under the warm bedcovers to read.

Tuesday — January 12, 1790

There is a special equation nobody knows how to solve. At first glance the equation appears very simple: $x^n + y^n = z^n$. However, not even the best mathematicians in the world know how to solve it. This is a Diophantine equation, named after Diophantus of Alexandria. There is no general method for solving such equations. What makes the equation more fascinating is a mysterious note that a mathematician wrote in the margin of a book before he died. His name was Pierre de Fermat, a Frenchman who lived about one hundred and thirty-five years ago.

Monsieur de Fermat claimed that the equation has no nonzero integer solutions for x, y, and z, when $n > 2$. After his death, his son found his father's *Arithmetica*, the book written by Diophantus hundreds of years ago—a book I am studying. The son discovered that in the margin of one page Fermat had written: "To divide a cube into two other cubes, a fourth power or in general any power whatever, into two powers of the same denomination above the second is impossible, and I have assuredly found an admirable proof of this, but the margin is too narrow to contain it."

Fermat meant that there are no whole number solutions for equations like these: $x^3 + y^3 = z^3$, $x^4 + y^4 = z^4$, $x^5 + y^5 = z^5$, and so on. These are equations of the general form $x^n + y^n = z^n$. I've solved $x^2 + y^2 = z^2$ many times. Using the Pythagorean Theorem to solve the triangle with two sides x and y equal to 3 and 4, respectively, yields the sum of perfect squares: $3^2 + 4^2 = 5^2$. Well, this equation is easy.

But Fermat claimed that when the exponent n is greater than 2, the equation $x^n + y^n = z^n$ has no solutions! It is very difficult to prove this statement because there are an infinite number of equations and an infinite number of possible values for x, y, and z. A great mathematician named Leonhard Euler obtained only a partial proof for the case $n = 3$. A full proof would require the inclusion of all cases to n infinite.

Would it not be wonderful if I could prove one day Monsieur Fermat's claim?

Thursday — January 21, 1790

Mother is so happy. She met Queen Marie-Antoinette at the Tuileries Palace. Not by herself, of course. Maman, Mme Fougeret, and other ladies in the committee of the *Charité Maternelle* had an audience with the queen. They asked for financial support to help poor unwed mothers and orphan children.

Maman is glowing with pride. She says that Her Majesty is a very compassionate and kind lady and agreed to support the charity. Maman thinks that the only reason some people blame the queen for the financial problems in France is because Marie-Antoinette likes to purchase fancy clothes and spends money on frivolous activities. Maman likes her very much and doesn't believe all the bad rumors against her.

As the daughter of the Empress Maria-Therese of Austria, Her Majesty was brought up believing her destiny was to become queen. And she did. At fourteen she married the crown prince of France. That was before I was born, in 1770. Four years later, she became queen of France when her husband was crowned King Louis-Auguste XVI.

Maman thinks that the stories of the queen's excesses are vastly overstated. Even Papa acknowledges that, rather than ignoring France's growing financial crisis, Marie-Antoinette reduced the royal household staff, eliminating many unnecessary positions that were based on privilege. Precisely because of this, the queen offended the nobles, adding their condemnation to the scandalous and false rumors spread by disloyal subjects. Maman insists that the aristocrats are the ones who reject the financial reforms the government ministers have proposed; the king favors social changes.

Now many political leaders are inciting the citizens against the monarchy. They will not appreciate the good deeds of Her Majesty. Maman, however, will always remember Queen Marie-Antoinette with affection. She spent the evening telling us how beautiful and gentle she is, how regal her demeanor, and how graciously she received the ladies in the Tuileries.

Wednesday — January 27, 1790

The best time for me to study is at night, after everybody goes to sleep. But twice a week I study in Papa's library when my parents leave to attend the salons of Mme de Maillard and Mme Geoffroy. The salons are intimate gatherings at which people meet and talk about philosophy, politics, literature, and many other topics of current interest. During those evenings my parents have supper there, and they do not return home until late at night. Thus, I can study in the library where it is warm and cozy with the fireplace that Mme Morrell keeps burning until Papa retires to bed.

Tonight is freezing cold, and instead of going out, my parents decided to play cards in the library, thwarting my plans to study there. Just now Millie came to bring me a cup of hot chocolate and whispered that everyone has gone to sleep. It means I can steal a little time for my calculations without interruptions.

What would I do without Millie? Especially now. Mother ordered Madame Morrell to stop stocking my heater after eleven o'clock, and to ration my candles. If Mother found out that Millie sneaks out in the middle of the night to bring me a candle or a cup of chocolate, the two of us would be in trouble. It is one thing to help me with the embroidery, but another to go against my mother's orders. I must search for the candles myself because it would not be fair to have Millie punished for helping me.

I am so glad I have my lap desk because I can write my notes bundled up under the blankets. The hot cup of chocolate keeps my hands warm, and I can write more legibly.

Sunday — February 7, 1790

I found a rare book I can't read because it is written in Latin. Papa explained that Latin is the universal language of scholars. They write in the language of the enceints so that their work is understood by others anywhere in the world. He added, "And you, *ma petite élève*, must know Latin if you wish to learn from this book." Papa knows I like challenges, and if he believes I can do it, then I will teach myself the language of mathematicians.

Even though I could not understand the words, I kept perusing the book until I found an equation that caught my attention. The equation looks simple, yet elegant and pure: $x^3 + 1 = 0$. So I copied it in my notebook assuming it'd be easy to solve. Then I discovered that there are three values of x that make the left side of the equation 0. It is clear that $x = -1$ is one solution since $(-1)^3 + 1 = -1 + 1 = 0$, but what are the other two solutions? It can't be $x = 1$ because I would get $(1)3 + 1 = 2$. And it can't be any number greater than 1 either, so what could the other two values of x be? I wish I could read Latin because maybe the solution is explained with words.

Oh, my candle is almost exhausted. It is freezing, and the embers in the heater are already cold. My hands feel like ice and I cannot hold the pen any longer. I better get some sleep and tomorrow I'll figure out how to solve this problem.

Sunday — February 14, 1790

Sir Francis Bacon said: "*Ipsa scientia potestas est,*" which means, "Knowledge itself is power." Bacon was an English philosopher who argued that the only knowledge of importance to man is empirically rooted in the natural world. Bacon believed that a clear system of scientific inquiry would assure man's mastery of the world.

Indeed, knowledge is power. I need to learn the language used by scholars to write about their discoveries. Of course, it is not easy to learn Latin by myself. I have to look up almost every word in the dictionary just to understand a bit of what I am reading. I guess the meaning of some words, but I have a hard time understanding an entire sentence. Translating takes so much time!

It occurred to me that I could take lessons from the nuns at the convent, since they are fluent in Latin. So I asked Maman, without revealing that it would help my study of mathematics. I expected that my mother would be most favourably disposed since this knowledge could be considered part of my religious education. Indeed, Maman promised to speak with the Mother Abbess at the Sisters of St. Joseph convent. I was delighted and couldn't wait to start my tutoring. But at dinner, Papa told me to wait until things settle down. He remarked that the Assembly just voted to suppress religious orders and monastic vows. Beginning today the government will give the nuns a choice of either leaving the convent and accepting government pensions, or remaining in designated monasteries chosen by the State. It's still too early to know what the Sisters at St. Joseph would do, so for now my lessons in Latin will have to wait.

Très bien. I'll continue learning Latin on my own until we find out. If it gets difficult, I'll tell myself *Nil desperandum*, which means "do not despair," and I will keep trying.

Saturday — February 20, 1790

I am studying equations that can be solved only with *imaginary numbers*. Imaginary numbers are the product of a real number and the square root of -1. An imaginary number is a multiple of a quantity called "i," which is defined by this property: i squared equals -1, or $i = \sqrt{-1}$. These numbers are called imaginary because they did not fit the definition of number that was available when first introduced. Imaginary numbers arose from the need to solve certain quadratic equations that many mathematicians believed had no solutions.

At first, I could not conceive such negative square roots. Then, I came up with a very simple way to see imaginary numbers and conclude that many problems in mathematics could not be solved without them. For example, if I have to solve the quadratic equation $x^2 + 1 = 0$, that means that $x^2 = -1$, thus the solution is $x = \sqrt{-1}$. Obviously, without imaginary numbers I would not know what $\sqrt{-1}$ is, and thus I could not arrive at the solution.

With the imaginary quantity i, and letting a and b be positive or negative real numbers, I can construct infinitely many numbers of the form $a + ib$ so that I can find the values of x that make my equation true; the two solutions to the equation $x^2 + 1 = 0$ are i and $-i$, since $x = -1$, or $x = \pm\sqrt{-1} = \pm i$. These solutions are complex imaginary. As I said, the imaginary unit is denoted and commonly referred to as i. Although there are two possible square roots of any number, the square roots of a negative number cannot be distinguished until one of the two is defined as the imaginary unit, at which point $+i$ and $-i$ are obvious. Since either choice is possible, there is no ambiguity in defining i as "the square root of -1."

Ancient mathematicians thought that it was impossible to take the square root of a negative number. That's because they had not thought of numbers that were negative when squared. The first mathematicians had only real numbers at their disposal, and thus didn't know how to take the square root of a negative number.

Imaginary numbers were invented a few hundred years ago. I say "invented" because all numbers are a creation of our minds. People create numbers to help them solve equations; numbers are not physical objects! So, mathematicians invented imaginary numbers because they needed them, just like they invented many other mathematical concepts.

I wonder what new mathematics will be revealed to me now that I know this number called i.

Friday — February 26, 1790

I am terribly upset with my sister Angélique. This afternoon, when I was in the library, she came into my bedroom and rummaged among my things. She opened the drawer in my lap desk and found my papers. Silly Angélique —when she saw my writing about imaginary numbers she ran to tell Mother that I was going crazy. She started teasing me, moving an imaginary pen in the air, saying, "O la la, Sophie plays with imaginary numbers!" Mother dashed in and demanded that I throw away my notes, saying that this was the reason she didn't approve of my study of mathematics. I tried to explain that imaginary numbers are not a figment of my imagination, but Maman wouldn't listen. She took all my notes and was about to burn them.

At that precise moment, Papa appeared and asked what the ruckus was about. After I told him, he smiled sympathetically and assured my mother I am not losing my mind. He persuaded her to return my notes. Maman did it reluctantly. First, she made me promise that I won't spend all my time in "this ridiculous pursuit unbecoming of a young lady." That is what my mother said, among other things I do not want to repeat. Why can't she understand that learning mathematics is not a silly pursuit? It is the only thing that matters to me!

My mother and my sister made such a fuss, thinking that my study of imaginary numbers is abnormal. Of course I am not crazy, but it would be futile to try to explain it to them. However, I should be patient and not be so hard on them for not understanding. It took centuries for these numbers to become fully understood and used.

It all began with the discovery of negative numbers. Before the seventeenth century, many mathematicians didn't know or accept negative numbers. Even great mathematicians like Blaise Pascal resisted this idea. Pascal was a French philosopher and mathematician, a colleague of Pierre de Fermat, who lived during the reign of Louis XIV. I read that Pascal once said: "Who doesn't know if you take 4 away from nothing, you're still left with nothing?" Pascal knew about negative numbers because, a century before, the Italian mathematicians Niccolò Fontana Tartaglia and Girolamo Cardano discovered negative roots when trying to solve cubic equations. These negative roots eventually led to the invention of imaginary numbers.

Cardano was solving cubic equations such as $x^3 = 15x + 4$, when he obtained an expression involving the square root of -121. Cardano thought that he could not take the square root of a negative number, however, he knew that $x = 4$ is one solution to the equation. But he was unsure about the other two solutions. He wrote to Tartaglia asking for his opinion on this puzzling answer, but Tartaglia was unable to help him. Cardano almost discovered imaginary numbers when he concluded that the problem of dividing 10 into two parts, so that their product is 40, would have to be $5 + \sqrt{(-15)}$ and $5 - \sqrt{(-15)}$.

Cardano wrote a book entitled *Ars Magna* (*The Great Art*) where he included negative solutions to equations, but he called them "fictitious" numbers. In his book, Cardano also noted an important fact connecting solutions of a cubic equation to its coefficients, namely, that the sum of the solutions is the negation of b, the coefficient of the x^2 term. A few years later, another Italian mathematician named Raphael Bombelli gave several examples involving these new numbers, and he gave the rules for their addition, subtraction, and multiplication.

René Descartes introduced the terms "imaginary" and "real" in his book *La Géométrie*. Descartes wrote: "Neither the true roots nor the false are always real; sometimes they are, however, imaginary; namely, whereas we can always imagine as many roots for each equation as I have predicted, there is still not always a quantity which corresponds to each root so imagined. Thus, while we may think of the equation $x^3 - 6x^2 + 13x - 10 = 0$ as having three roots, yet there is just one real root, which is 2, and the other two, however, increased, diminished, or multiplied them as we just assigned, remain always imaginary." In 1777 Leonhard Euler recommended the general use of imaginary numbers. Chapter XIII of his book *Elements of Algebra* was devoted to imaginary numbers. Euler wrote that the calculation of imaginary quantities is of the greatest importance and gave several examples to illustrate how to extract the root of a negative number.

I solved the cubic equation $x^3 - 6x^2 + 13x - 10 = 0$ that Descartes mentioned. In addition to the solution $x = 2$, the other two roots are $x = 2 + i$ and $x = 2 - i$. Without imaginary numbers, I could not have defined the square root of the negative numbers. How wonderful! It seems as if a new door to mathematics is open to me now that I know imaginary numbers.

Tuesday — March 2, 1790

My father asked me how to compound interest. He lends money and, to compute the interest for each loan, he uses a simple formula $I = P \cdot R$, where $I =$ interest, $P =$ principal or original amount loaned, and $R =$ rate of interest. Now Papa wants to compound interest to charge not only for the original loan, but also charge for some of the interest owed. He wants to charge interest on the interest because some people take too long to pay him back.

I found the formula $A = P(1 + R/n)^{nt}$ to calculate the compound interest, where A is the total amount he would be paid, t is the elapsed time, and n is the number of times he can compound the interest in the time t. Papa wants to figure out how many times he can compound the interest in order to get the highest return for his loans.

Since Papa did not tell me the interest he charges for each loan, I assume it is 100% per year, and further assume he loans one livre (1 l). These values might not be realistic, but this assumption will simplify the analysis.

Thus, with $P = 1\,l$ and $R = 1.00$, and $t = 1$, I get the following:

Compounding annually (once a year),
$$n = 1 \quad \text{and} \quad A = 1\,l(1 + 1.00/1)^{1 \times 1} = 2.0\,l$$

Compounding biannually (twice a year),
$$n = 2 \quad \text{and} \quad A = 1\,l(1 + 1.00/2)^{1 \times 2} = 2.25\,l$$

Compounding quarterly (four times a year),

$$n = 4 \quad \text{and} \quad A = 1\,l(1 + 1.00/4)^{1 \times 4} = 2.44\,l$$

Compounding every month,

$$n = 12 \quad \text{and} \quad A = 1\,l(1 + 1.00/12)^{1 \times 12} = 2.61\,l$$

I keep increasing n, doubling and tripling it, but A grows at a very slow pace. Even if the interest is compounded daily, I do not see much growth.

There appears to be a limit to the earnings from the compound interest formula. I have to tell Papa that he could compound the interest every day, but the most he would net is limited. So, even if the interest were compounded every hour, he still would get less than three times the amount he lends. Amazing! I wonder why?

Sunday — March 7, 1790

The game of chess is intimately connected to mathematics. Today I learned how. I was in the library, playing chess by myself, when Papa arrived with Monsieur de Maillard. He startled me when he moved a piece on the board and commented on my strategy. As they settled in armchairs to drink coffee, M. de Maillard told me an extraordinary tale of the origin of this game. I'll attempt to recount the story.

Many centuries ago, there was a king who wanted a unique game that no one else had, a game that was so original that it could be played over and over in endless combinations, a game that would teach his children to become better thinkers and better leaders on the battlefield. With this in mind, a wise man invented the game that we now call chess. The monarch was very pleased, and he offered as a payment anything the man would want—gold, jewels, anything. The wise man asked to be paid with wheat grain. When asked how much, the man replied that he wanted the amount based on the number of squares on the game board.

His formula was simple; for square one, he wanted one grain, and to double the previous amount in each subsequent square of the chessboard. In other words, for square two, it would be two pieces (1 times 2), for square three, it would be four pieces (2 times 2), for square four it would be eight pieces (4 times 2), and so forth until all sixty-four squares were used in this proportion to determine the total amount of grain. The king was surprised by the seemingly modest request, but ordered a sack of wheat grain to be brought. The servants patiently began to place the pieces of wheat on a bucket calculating the number using the formula requested by the wise man. But the grain ran out and more sacks were brought up. Soon, they discovered that not even all the grain in the kingdom would be sufficient to give the man the amount that corresponded to half of the squares on the chessboard.

M. de Maillard asked me if I could determine how much grain would be needed. I didn't know offhand, but I scurried to my room and decided to give it a try. I started with numbers to illustrate the distribution of wheat grain on the chessboard. Since there are eight squares on the first row, I write:

1	2	4	8	16	32	64	128

in the second row:

256	512	1024	2048	4096	8192	16,384	32,768

in the third row:

65,536   131,072   262,144   ....

Très bien! Enough!

I have to stop here. Now I just imagine the huge numbers on the squares of the fourth row and the remaining squares on the chessboard. Without doing the calculation I know that on each square I can represent the number of pieces by the number 2^n, with the first square $2^0 = 1$, and the number on the last square $2^n = 2^{63}$. Then I have to add the number of pieces on all squares, making the total number enormous. Even 2^{63} is such a huge number that I cannot visualize it!

How can I tell this story using only mathematics? I could write a sum $S = 1 + 2 + 4 + 8 + 16 + 32 + \cdots + 2^{63}$, which gives me an idea of the total number of pieces.

That's all I can think right now.

Monday — March 15, 1790

Today was cold and gray, and the rain made me feel gloomy. After dinner, we gathered in Papa's library around the cozy fireplace. Angélique was restless and cranky so I offered to teach her how to play chess. At first she was not that interested, but then I told her about the monarch who wanted a game to teach his children to be smarter, and she sat listening quietly. My sister likes fairy tales, so it was easy to embellish the story by creating more characters that she found interesting, although none of them had anything to do with chess—or with mathematics!

1	2^1	2^2	2^3	2^4	2^5	2^6	2^7
2^8	2^9	2^{10}	2^{11}	2^{12}	2^{13}	2^{14}	2^{15}
2^{16}	2^{17}	2^{18}	2^{19}	2^{20}	2^{21}	2^{22}	2^{23}
					2^{61}	2^{62}	2^{63}

In my tale the wise man became a handsome prince; he would marry the king's beautiful daughter after winning the game. I also added that the ruler was very pleased with the new game, and thus offered, as a dowry for the lovely princess, anything the smart prince would want—gold or diamonds. I also changed the part of the story where the wise man asked to be paid with wheat grain. To keep Angélique's interest, I changed the story slightly by saying that the prince asked for diamonds! My sister was enchanted and was more interested in learning to play, imagining perhaps the sparkling diamonds on each square of the chessboard.

We spent some time playing chess, and I explained the method to determine the number of grains (or diamonds) that would result from doubling the number in each subsequent square of the chessboard. I don't think Angélique grasped what I was telling her, but she listened patiently. The number of grains on the chessboard would be a sum of 2s raised to increasing powers, from $n = 0$ to $n = 63$: $S = 2^0+2^1+2^2+2^3+2^4+2^5+\cdots+2^{63}$. More compactly, $S = 2^{64} - 1$ grains.

Hence, just counting the grains that correspond to the first ten squares, it would yield a sum of

$$S = 2^0 + 2^1 + 2^2 + 2^3 + 2^4 + 2^5 + 2^6 + 2^7 + 2^8 + 2^9$$
$$= 1 + 2 + 4 + 8 + 16 + 32 + 64 + 128 + 256 + 512$$
$$= 1023.$$

That would be a lot of wheat indeed!

Thursday — March 25, 1790

Mother does not appreciate me! She doesn't understand why I study mathematics. She can't! This evening I overheard her telling Papa that I am taciturn and self-absorbed. She complained that there are times when she asks me a question, and I do not even lift my head up from the book I am reading. With an exasperated voice Mother confided that, "at times Sophie sits there with her eyes glazed over, looking away like in a trance." I do not know what Papa replied after that. I did not want to hear anymore so I left as quietly as I had arrived.

I am hurt and upset. Why does my mother call me self-absorbed? Does she not understand that mathematics requires deep thinking and full concentration? When I am engaged in the solution of a problem, or the study of a new topic, I submerge myself in it, leaving behind everything else. This should not worry her.

My dear mother; she believes in the idea of feminine fulfillment advocated by the philosopher Jean-Jacques Rousseau, that women should devote themselves to the happiness of their husbands and children. Maman thinks that women should remain silent and not express their opinions, especially if they are in conflict with their husband's point of view. She believes that a girl should learn domestic skills to become a good wife and mother, not a scholar. I disagree but cannot contradict her. I just wish she'd accept me for what I am.

Holy Thursday — April 1, 1790

We went to church and witnessed something amazing. For the first time in history King Louis and Queen Marie-Antoinette performed the *Pedilavium*—washing of the feet. The ceremony began when twelve of the poorest people, dressed in new clothes donated by the monarchs, walked to the front of the church. They sat on a bench with their right feet bare and rested them on the edge of a basin of water. His Majesty approached them and "washed" each person's foot by pouring water over it with a scoop. The queen followed, placing a white napkin over the wet foot. It was all very solemn; people watched with admiration and awe because this is the first time a king of France has performed such a humble act. It's the king's way to gain his subjects back and ease the social tensions of the past year.

I turned fourteen today. We spent the day in solitude and prayer. On this day people remember the trial of Jesus Christ that led to his crucifixion. During this holy week we fast and pray in preparation for Easter. Fasting is not difficult for me because it just means that we eat very little every day, mainly bread and water, but my little sister Angélique still struggles with it, for she can't abstain from eating desserts. By evening, Madame Morrell fixed her some bouillon because she was starving. Later, Millie gave her cheese and a bit of jam. Before going to bed Angélique asked me whether God would be mad at her for not fasting.

I hear Mme Morrell extinguishing the lamps in the hall and Millie climbing the stairs to their quarters. The midnight bells will toll very soon so it's time to snuff my candle.

Easter Sunday – April 4, 1790

Today was cold, but the sun was out, shining brightly over the clear blue sky. After Easter Mass we rode to the Tuileries gardens. The pathways in the gardens were crowded; women were parading with their newest dresses and feathered hats, celebrating the new beginning that Easter represents.

As we strolled by the Tuileries main entrance, we recognized the new Dauphin, the heir to the throne, playing with his sister and their governess. The boy must be five years old now, very friendly and precocious. The princess, a blond girl who is about my little sister's age, was aloof and hardly looked at the people. Angélique waved at the royal children, and the Dauphin waved back enthusiastically. My sister was thrilled, assuming that he had noticed just her, but the boy was waving at all the people that were watching him play in the palace's garden. Maman says that the Tuileries is not as sumptuous as Versailles, so the royal family had to adapt to a most modest home. The little boy, however, did not seem to mind. He looked as happy as any other child his age.

Papa is going to purchase some land located in the north side of Paris. He will buy it at the auction of properties that belonged once to the Church of France. Now the government is selling it in order to pay for the national debt. Papa asked me to estimate how much land he can buy with 2000 livres. This is quite a trivial problem, mathematically. But for somebody trying to buy property it is very important.

For example, if the lot were perfectly square, its area would be the product of the length of one side by itself, that is, $x \cdot x = x^2$. With x^2 representing an area, I write the relationship between the area and the cost. For example:

$x^2 = 2000$. Solving this equation, I get the length of the lot, $x = $ square root of $2000 \approx 44.7$ m. With 2000 livres Papa can buy a square lot that measures 44.7×44.7 meters. Now, if Papa also has to buy a fence to enclose the land, the formula must include both area and length. The length of fence is simply the perimeter of the square lot, or $4 \cdot x$. Now I assume that Papa has 2021 livres to purchase everything. In this case, I write a relation $x^2 + 4x = 2021$. To solve this quadratic equation I note that $(x^2 + 4x)$ represents the first two factors of the binomial $(x + 2)^2$, which expanded is $(x + 2)^2 = x^2 + 4x + 4$. Thus, I can use this fact to write my equation as

$$x^2 + 4x + 4 = 2021 + 4$$

(I just added a 4 to both sides of the equation). Simplifying further, I get

$$x^2 + 4x + 4 = 2025$$
$$(x + 2)^2 = 2025.$$

Très bien. Now, what if the land is not perfectly square? I could assume, for example, that the lot is rectangular. In such a case, the area is equal to $x \cdot y$, and the perimeter is equal to $(2x + 2y)$. My equation would then be: $x \cdot y + 2(x + y) = 2021$. How can I solve it? Easy! All I need is a relationship between x and y. Let's say that $y = 3x$, then I can substitute as follows:

$$x \cdot y + 2(x + y) = x(3x) + 2\big[x + (3x)\big] = 3x^2 + 2x + 6x = 2021,$$
$$3x^2 + 8x - 2021 = 0.$$

Once again, I have a quadratic equation that is easy to solve. So, no matter what geometric shape I am given, as long as I can find a relationship between the sides x and y, I will arrive at a solution.

Friday — April 9, 1790

Today I received an unexpected and wonderful gift! I can hardly contain myself to write about how it happened. We were sitting in Mother's study, ready for our writing lessons, when Millie opened the door followed by a stranger. The old gentleman introduced himself as the valet of Monsieur the Marquis de Condorcet and announced gravely: "Letter for Mademoiselle Marie-Sophie Germain." I was unsure. Did he call my name?

All eyes turned to me at once, stumped as I was. After clearing his throat, the gentleman repeated, "I bring this communication for Mlle Sophie Germain on behalf of the Marquis." With a nod, Maman encouraged me to stand up and accept the letter, which the old man handed to me along with a package. I was stunned. I rose, I stammered a few words that were not heard, and took the parcel. I am sure my checks were red, inflamed by both my timidity and my delight, and I managed to curtsy and thank him. My sisters rushed to me after the man left, and Maman urged me to read the missive and to open the package. They were excited, but not as much as I was.

There was no mistake; the letter was addressed to me! My name was written elegantly in black ink letters on the ivory parchment. I broke the aristocratic wax seal with trembling hands. The message was short: *May you find Euler's memoir stimulating. It is for the few who truly appreciate the beauty of mathematics.* Signed "Jean-Antoine-Nicolas Caritat, Marquis de Condorcet." My hands trembled in anticipation, and Angélique had to contain herself from removing the wrapping for me. The package contained two volumes with leather covers of the book *Introductio in analysin infinitorum* by Leonhard Euler. Angélique quickly lost interest when she saw mathematics and the text written in Latin.

Maman was unsure that I should accept a gift from such a distinguished gentleman. She is pleased with this honour, of course, but she does not understand why a man would encourage my study of mathematics. She cannot prevent me from reading these books, but after dinner she reminded me to go to bed early. Which I do, and she knows it. Maman also made me promise to write a nice letter to the Marquis to thank him for his generosity.

I am thrilled to receive this unexpected but most extraordinary gift. Papa introduced me to the Marquis when we were at the bookshop near the Louvre. While exchanging pleasantries, Papa mentioned my interest in mathematics. M. Condorcet beamed, adding that he, too, is interested in the most beautiful of sciences. My heart skipped a beat when I heard him speak like that. After we said goodbye, Papa praised M. Condorcet, stating he is a man of considerable erudition, a great mathematician at the Academy of Sciences.

M. Condorcet must think I need to study this book. Even if I do not understand every sentence I read, I can study the equations. So far, I translated the title of the book as "Introduction to Analysis of the Infinite." M. Condorcet's excellent gift means more to me than all the gold in the world. I must find the words to express my immense gratitude in a letter.

My books on the analysis of the infinite wait on my table, beckoning me to begin the journey of discovery and learning.

Friday — April 16, 1790

Introduction to Analysis of the Infinite, which Euler published in 1748, opens with the definition of *functio*, a Latin word that means "function." Euler defined a function as a variable quantity that is dependent upon another quantity. For example, the analytic expression $y = 2x^2$ is a function, written as $f(x) = 2x^2$. Thus, a function is simply an association between two or more variables; to every value of each independent variable it corresponds to exactly one value of the dependent variable, in a specified domain where the function is valid. When I write the expression $f(x) = 2x^2 + 7$, for each value of x I get a value of the function $f(x)$; so $f(x)$ denotes the dependent variable, and x is the independent variable.

An equation in which the dependent and independent variables occur on one side of the equality sign is said to be in implicit form. For example, the expression $2x^2 - 2y = 6$ is written in implicit form, but it can be written in the more clear explicit form, $y = x^2 - 3$, or also as $f(x) = x^2 - 3$.

The domain of the function is the collection of all values that the independent variable can take. For example, given the function $f(x) = 1 - x^2$, and defining the independent variable x as a real number, the domain of this function must be the collection of all real numbers from $-\infty$ to $+\infty$, where the symbol ∞ represents "infinity" and the minus and plus signs mean that infinity can be negative or positive. This is because the numerical value of x^2 will always be zero or positive. As x goes toward ∞ or $-\infty$, $f(x) \to -\infty$. Of course, $f(x)$ will reach its greatest value when $x = 0$. The range of this function includes the number 1, and all the real numbers less than 1. Written with mathematical notation, the range is given by $-\infty < f(x) \le 1$.

There are many types of function. Linear functions are those defined by a linear equation such as $f(x) = ax + b$. Power functions are those defined by a number raised to a power, $f(x) = a^n$. Quadratic functions, just like the name implies, are defined by a quadratic polynomial, $f(x) = ax^2 + bx + c$. Exponential functions are of two forms, $f(x) = a^x$ to the base a, where $a > 0$, and, $f(x) = e^x$ to the base e, where e is a number defined by Euler. He also uses the trigonometric functions $f(x) = \sin x$, and $f(x) = \cos x$.

Euler defines polynomials and rational functions. He gives the general form for a polynomial as $a + bz + cz^2 + dz^3 + ez^4 + fz^5 + \&c$, but he doesn't explain the meaning of the "&c" (does *et cetera* mean a few more terms or infinity?). Euler does not specify if the coefficients are real or imaginary. I wonder....

Saturday — April 24, 1790

We spent the afternoon at the *Jardin des Tuileries*. Maman and Madeleine were eager to see the latest fashions. Parisian women have this silly way of showing off their newest gowns by pretending to go for a stroll through the gardens. I like to go only because Maman allows me to visit the bookstores nearby. There is one shop I especially like that is filled with new and used books at very reasonable prices. I found Jean de la Fontaine's *Fables*, very pretty with colorful illustrations, much nicer than the old book at home. So I bought it for Angélique.

My study of Euler's book is progressing slowly, as I have to translate the text to understand his analysis, which is a bit challenging. He introduces two mathematical terms: sequence and series. It took several days of study, but now I understand what they mean.

A *sequence* is a set of numbers arranged in an orderly fashion such that the preceding and following numbers are completely specified. The numbers in a sequence are called "terms." The "general term" defines the rule of the sequence. For example, if n is the ordinal number of a term in the sequence, 1 is the first term, and the general term is $2n - 1$, the sequence is written as $1, 3, 5, 7, \ldots, 2n - 1$.

A *series* is the sum of the terms of a sequence, and it can be a finite or infinite series. Finite sequences and series have defined first and last terms. For example:

$$\sum_{n=0}^{5} \left(\frac{1}{2}\right)^n = 1 + \frac{1}{2} + \frac{1}{4} + \frac{1}{8} + \frac{1}{16} + \frac{1}{32}$$

where n takes values from 0 to 5. The symbol $\sum$ represents a sum.

Infinite sequences and series continue indefinitely (to ∞). For example the series:

$$\sum_{n=1}^{\infty} \left(\frac{1}{n}\right)^n = 1 + \frac{1}{2} + \frac{1}{3} + \frac{1}{4} + \cdots$$

where n takes values from 1 to ∞.

Euler includes derivations of the power series for the exponential, logarithmic, and trigonometric functions; the factorizations of the sine and cosine functions; and the consequent evaluation of the "sum of the reciprocal squares." Euler defines logarithms as exponents and the trigonometric functions as ratios.

There is a new mathematical expression that Euler used to manipulate the infinite series. First, how do I find the sum of an infinite series? I must find its limit. For example, if I add more terms to a convergent series, its terms become smaller and smaller. In other words, the terms of a convergent series tend to 0. The sum of such a convergent series is referred to as "the sum to infinity," and mathematicians simply write this as $\lim_{n \to \infty} S_n$, where lim stands for "limit," a term from the Latin word *limes*. The mathematical meaning is still a little confusing, but I'm sure the next sections of Euler's derivation will help me understand what the limit of a function is and how to determine it.

Friday — April 30, 1790

I continue studying Euler's book. The two volumes of the *Introductio in analysin infinitorum* cover a wide variety of subjects, including the infinite series and expansion of trigonometric functions. Right now I am studying a new number that Euler calls "*e*." It is a most amazing number that Euler connects to the trigonometric functions in a most intriguing manner. Euler begins by defining the number e with an infinite series expansion:

$$e = 1 + \frac{1}{1} + \frac{1}{1 \cdot 2} + \frac{1}{1 \cdot 2 \cdot 3} + \frac{1}{1 \cdot 2 \cdot 3 \cdot 4} + \cdots.$$

Euler shows that the number e is the limit of $(1 + 1/n)^n$ as n tends to infinity. I still do not know the meaning of this number e, but I imagine it must be similar to π. I think so because, by definition, e is equal to an infinite series. This implies that its value is not exact, since one can always add another term to the series and keep on adding to infinity.

In §122, Euler gave an approximation for the number e to 18 decimal places! He wrote $e = 2.718281828459045235360287$, without saying where the value came from. Perhaps Euler calculated this number himself, but he did not say how. However, I computed the values of the first seven terms in the infinite series to verify the value of e by myself:

$$1 = 1.0000000$$

$$\frac{1}{1} = 1.0000000$$

$$\frac{1}{1 \cdot 2} = \frac{1}{2} = 0.5000000$$

$$\frac{1}{1 \cdot 2 \cdot 3} = \frac{1}{6} = 0.1666666$$

$$\frac{1}{1 \cdot 2 \cdot 3 \cdot 4} = \frac{1}{24} = 0.0416666$$

$$\frac{1}{1 \cdot 2 \cdot 3 \cdot 4 \cdot 5} = \frac{1}{120} = 0.008333$$

$$\frac{1}{1 \cdot 2 \cdot 3 \cdot 4 \cdot 5 \cdot 6} = \frac{1}{720} = 0.0013888.$$

Adding them, I get 2.7180553, a number that is near the value for e given by Euler. So, if I could add all the terms of the series, perhaps I could prove that

$$e = 1 + \frac{1}{1} + \frac{1}{1 \cdot 2} + \frac{1}{1 \cdot 2 \cdot 3} + \frac{1}{1 \cdot 2 \cdot 3 \cdot 4} + \cdots = 2.718281\ldots.$$

Euler also wrote the fraction expansion of e, and noted a pattern in the expansion. He did not give a proof that the patterns in the series continue, but I am sure that it should lead to a proof that e is irrational. Whatever the number e means, it has to be irrational, just like π. And if π is related to the circle, what is e related to?

There are other very interesting results in Euler's book. Tomorrow I will attempt to translate the chapter titled *De partitione numerorum* (I think it means "On the Partition of Numbers", but can't be sure).

Monday — May 10, 1790

It would be delightful to share with someone what I am learning from Euler. He derives so many beautiful relations, so simple and yet so intriguing. It's amazing how the mind of one person can create such beauty with numbers. One example of his genius is this elegant expression, which I call Euler's equation:

$$e^{i\pi} + 1 = 0.$$

It connects five unique fundamental numbers in a relation of exquisite simplicity: the principal whole numbers 1 and 0, the chief mathematical signs $+$ and $=$, and the special numbers, e, i, and π. I do not know what this relationship means, but I can only imagine the secrets that this beautiful equation guards within.

Euler's exquisite identity is a special case of $e^{ix} = \cos x + i \sin x$, a formula from which, as he noted, complex exponentials can be expressed by real sines and cosines, my favorite functions. He deduced the special relation using De Moivre's theorem, which states that for any real number x and any integer n, the sine and cosine are related in this form:

$$(\cos x + i \sin x)^n = \cos(nx) + i \sin(nx).$$

This relation is important because it connects complex numbers and trigonometry. I do not know what its significance is, so I must attempt to follow Euler's analysis to understand how he developed his equation for e. I start with the series expansion of the three functions e^x, $\sin(x)$, and $\cos(x)$:

$$e^x = 1 + x + \frac{x^2}{2!} + \frac{x^3}{3!} + \cdots$$

$$\sin(x) = x - \frac{x^3}{3!} + \frac{x^5}{5!} - \cdots$$

$$\cos(x) = 1 - \frac{x^2}{2!} + \frac{x^4}{4!} - \cdots.$$

Then I write:

$$e^{ix} = 1 + ix - \frac{x^2}{2!} - i\frac{x^3}{3!} + \frac{x^4}{4!} + \cdots$$

$$= \left(1 - \frac{x^2}{2!} + \frac{x^4}{4!} - \frac{x^6}{6!} + \cdots\right) + i\left(x - \frac{x^3}{3!} + \frac{x^5}{5!} - \frac{x^7}{7!} + \cdots\right)$$

$$= \cos x + i \sin x.$$

The only way to obtain Euler's equation is by changing the variable, that is, letting $x = \pi$. Then I can write:

$$e^{i\pi} = \cos \pi + i \sin \pi = -1 + i \cdot 0 = -1$$

which means that $e^{i\pi} = -1$, or

$$e^{i\pi} + 1 = 0.$$

Voilà! It's beautiful!—if I didn't know that I could substitute the variable x with the constant π, I probably would not have arrived at the expression for $e^{i\pi}$. Euler did not explain why he made the substitution, or why it is valid. But it works! That's another example of the way mathematicians think. It requires one to know a lot, and to "see" the right connections between equations and numbers, just as Euler did.

Saturday — May 15, 1790

Now I understand this equation: $e^{ix} = \cos x + i \sin x$. First of all, the equality implies that e^{ix} is a complex number with two components, a real part (cos x) and an imaginary part (sin x). Furthermore, if $x = \pi$ (or any multiple of π), the value of the exponential function depends on the cyclic value of the trigonometric functions. For example, with $x = \pi$, it yields $\cos \pi = -1$ and $\sin \pi = 0$; then I can write $e^{i\pi} = \cos \pi + i \sin \pi = -1 + i \cdot 0 = -1$, which leads me directly to Euler's relation $e^{i\pi} + 1 = 0$. Euler knew his trigonometry!

And if $x = 0$, $\cos 0 = 1$ and $\sin 0 = 0$, I get: $e^{i0} = \cos 0 + i \sin 0 = 1 + i \cdot 0 = 1 = e^0$. This shows that, indeed, the number e follows the same rules of exponents that I learned in algebra: any number raised to the 0-power is equal to 1.

So, what happens if the imaginary unit i is raised to the ith power? What kind of number is i^i? Is it real or imaginary? I will find out.

Start with Euler's equation $e^{i\pi} = -1$, which is the same as $e^{i\pi} = i^2$ (since by definition $i = \sqrt{-1}$ or $i^2 = -1$). If I raise both sides of this equation to the i-th power, I get,

$$\left(e^{i\pi}\right)^i = \left(i^2\right)^i$$

or

$$e^{i(i\pi)} = e^{-\pi} = \left(i^2\right)^i = \left(i^i\right)^2.$$

Now I take the square root of each side,

$$\left(e^{-\pi}\right)^{1/2} = \left[\left(i^i\right)^2\right]^{1/2} = i^i,$$

which is the same as

$$\frac{1}{(e^\pi)^{1/2}} = i^i.$$

This means that the imaginary power of the imaginary unit is a real number!

Saturday – May 22, 1790

What is infinite? The other night, observing the twinkling stars in the dark sky, I began to wonder about the size of the universe. There are so many stars in the firmament, so far away. Does the cosmos have a limit? Since the universe is made up of matter, it probably has bounds, otherwise there would have to be an infinite amount of matter.

What about the numbers—are they infinite? Yes! Of this I am sure. Because no matter how large a number is, I can always add another one and make it even larger. How large is "large"? If I use an exponent to represent a very large number, say 10^n where n is a number greater than any number I could count, I cannot even conceive in my mind the number of digits it would have. Because even if I got tired of counting, I could always add one more number, that is $10^n + 1$, so this number is greater than my original largest number, $10^n + 1 > 10^n$.

Mathematicians define *infinite* as an unbounded quantity greater than every real number. The real numbers include all the integers, rational, and irrational numbers. Thus, whatever number I choose to be the last in my mind, I could as easily make it greater by adding 1, and so I could never reach the last number. Furthermore, since both 10^n and $10^n + 1$ are already very large numbers, they are infinitely large. I imagine this is like adding a drop of water in the ocean; the drop makes no difference to the size of the ocean. Therefore, adding one to infinity should yield infinity!

Sunday — June 6, 1790

On Thursday we celebrated the feast of Corpus Christi at St-Germain church. We went there because Maman wanted to see the king and queen, who led the procession of the Blessed Sacrament. After all the riots and protests of the past year, the monarchs were gracious to join the people in this manner.

Mother would be mortified and angry with me if she knew that during the Mass, my mind was wandering, distracted by mathematics. I was thinking about two very simple equations that appear to be related; yet each one led me into a different realm of mathematics.

The equations are of the form $x + a = 0$ and $x^2 + a = 0$, where a is any positive number. These two seemingly innocent equations require solutions that are outside the confines of real numbers. The solution of equations such as $x + a = 0$ requires negative numbers. For example, $x + 1 = 0$, has the solution $x = -1$. On the other hand, the solution to $x^2 + a = 0$ is an imaginary number. As I discovered, the number $\sqrt{-1}$, which is called the imaginary unit, is defined as one of the solutions of the quadratic equation $x^2 + 1 = 0$ (the other solution is $-\sqrt{-1}$).

What this means is that the two equations $x + a = 0$ and $x^2 + a = 0$ have no solutions unless we introduce negative and imaginary numbers, respectively. That is why I find mathematics so fascinating. To solve certain equations, even simple ones, sometimes one must develop new mathematics if no solution is possible within the realm of what is already known.

Does this mean that all problems in the universe have solutions? Well, if we identify a problem, indubitably it must have a solution. That's why my dream is to learn mathematics, and devote my life to seeking out answers to difficult questions.

Saturday — June 12, 1790

Ever since I caught sight of a colorful hot-air balloon floating over Paris, I've dreamed of flying among the clouds. That is just a fantasy, of course. Mother would never allow it. But it is wonderful to imagine that I could fly. When I was little, Papa read me the myth of Daedalus and Icarus, who wanted to imitate the grace and ease of flying birds. And then I dreamt that I, too, had wings.

I've read about Leonardo da Vinci and his inventions. He studied the flight of birds and sketched flying machines in a world that only he could see. Leonardo wrote: "If man could subjugate the air and rise up into it on large wings of his own making." Oh yes, I know that the human body is not meant to fly like birds, but wouldn't it be wonderful to at least fly aboard a flying machine like the ones in Da Vinci's imagination?

We went to the *Jardin des Tuileries* and then stopped at the Palais-Royal. It was crowded and noisy; it seemed as if all Parisians had agreed to meet there today. There is always something new to catch people's attention; it's like going to a festival. We saw acrobats, magicians performing all sorts of astonishing tricks, and vendors selling amusing toys. There were people dressed in multicolored costumes, dancing and singing in foreign languages. The arched passages were full of ladies wearing beautiful gowns and the most outrageous hats I've ever seen. Mother and Madeleine also visited the *Camp des Tartares*, mainly to browse the shops and boutiques, looking for fashion ideas.

After shopping, Madeleine met with her fiancé at the Café de Foy. I waited with them while my parents took Angélique to the marionette theater. We sat at a table in the central garden, and M. Lherbette ordered refreshing lemonade drinks. Later Papa took me to a special bookshop that sells scientific books from all over the world.

Papa gave me money to purchase a translation from one of Galileo's greatest works, the one that deals with his astronomical discoveries. The book is titled *Sidereus Nuncius*, which can be translated as "The Starry Messenger," and describes what he observed with his telescope. The original manuscript was published in Venice in 1610. It is fascinating; Galileo claimed there are mountains on the moon! He saw through his instrument many more stars than can be seen with our eyes, and above all "four planets swiftly revolving about Jupiter." Galileo was referring to his observation of Jupiter's moons. This is the perfect book to read during the warm nights of summer when the moon shines adorned by the twinkling stars. I must stop writing now to resume reading *The Starry Messenger*.

Thursday — June 17, 1790

It struck me like lightning on a dark night. There it was, the connection that I had not seen until now, the relationship that exists between logarithms and the number e. By definition, e is the limit value of the expression $(1 + 1/n)$ raised to the nth power, when n increases indefinitely. Seen this way, I did not connect it to logarithms. I had taken logarithms simply as tools to write a very large number in compact form. I knew that if three real numbers a, x, and y are related as $x = a^y$, then y is defined as the base-a logarithm of x. That is, $\log_a x = y$. For example, $1000 = 10^3$, therefore I can write $\log_{10} 1000 = 3$. This means that 3 is equal to the logarithm of 1000 when the base is 10.

Euler had the mind of a genius, as he discovered such incredible relationships. I am so glad to see the connection, too.

Friday — June 25, 1790

All French citizens are socially equals. Or at least in principle. On Saturday the Assembly voted to abolish the nobility, including the removal of aristocratic titles. That means that Monsieur the Marquis de Condorcet will now be addressed simply as Monsieur Condorcet, and the Duchess de Quelen will be, from now on, just Madame de Quelen. But removing their title—does it make aristocrats equal to the peasants in the countryside, or the workers in the factory? I do not think so. I don't believe that it is possible to achieve true social equality. Some people will always have more wealth and power than others; how can the poor become equals with them?

However, who needs aristocratic titles when one is an educated person? I am more impressed by scholars, by people who are recognized by their intelligence and their erudition. I'd like to believe that the Marquis de Condorcet would rather be called "monsieur the mathematician" than "monsieur the marquis."

I am fascinated and genuinely inspired by the illustrious mathematician Leonhard Euler. He did not have an aristocratic title, but he achieved immortality! He was born on April 15, 1707 in Basel, Switzerland. When he was fourteen, he entered the University of Basel; three years later, he was granted a master's degree in philosophy. His father expected him to become a minister, but Euler liked mathematics better. His teacher, Jean Bernoulli, who was a famous mathematician, convinced Euler's father to let him study mathematics. And that is how Euler became a mathematician. At nineteen he submitted two research memoirs to the French Academy of Sciences, one on the masting of ships, and the other on the philosophy of sound. In 1730, Euler became professor of physics at the Russian Academy of Sciences in St. Petersburg, and three years later he became professor of mathematics. Euler married the daughter of a Swiss painter and had thirteen children.

Euler had an extraordinary memory. I read that once Euler did a calculation in his head to settle an argument between students whose computations differed in the fiftieth decimal place. Euler was a very prolific mathematical writer and published hundreds of scientific papers. He won the prize from the French Academy of Sciences twelve times!

Euler suffered through illness and personal misfortunes. Euler lost sight in his right eye when he was relatively young and became completely blind many years later. His unquestioning faith enabled him to accept blindness with courage. Aided by his amazing memory, and having practiced writing on a large slate when his sight was failing him, Euler continued to publish his results by dictating them to his pupils. In fact, as an aging and completely blind scholar, Euler dictated his *Algebra* book to an assistant.

Leonhard Euler not only made advancements in mathematics, but also in astronomy, mechanics, optics, and acoustics. Euler was also first to prove that e is irrational. I could go on and on writing about Euler's many achievements. This inspires me more to learn mathematics and aspire to become a little bit like him.

Tuesday — July 6, 1790

Are women equal to men? Are we now allowed to attend the university like boys do? Well, according to Papa, today is the beginning of a something that will benefit women. I'm so euphoric that I am getting ahead of myself. Let's start from the beginning. I was in Papa's study this evening when his friends came to their regular Tuesday meeting. As I tried to leave, Papa asked me to stay. I didn't know why it was important since I never participate in their discussions. Soon it became clear. When all of his friends were comfortably seated, my father pulled out of his desk a pamphlet that he waved in his hand. "This," Papa said excitedly, "this is what France needs—complete civil equality for women and men!"

Papa explained that the Marquis de Condorcet had just published an essay arguing in favor of granting the same rights of citizenship to women. I was stunned. The gentleman who had taken seriously my interest and desire to learn mathematics is indeed the enlightened man I had assumed he would be.

Papa's friends, however, had different points of view on the matter. Some argued that "women have never been guided by reason and, therefore, they could not possibly use good judgment" in matters related to political and civil duties. Others even claimed that men have superiority of mind. However, my father refuted those silly arguments, and the discussion became very heated. At that moment, Papa placed M. Condorcet's pamphlet in my hand, and I dashed off with it. I came to my room to read it.

M. Condorcet's essay is titled *On the Admission of Women to the Rights of Citizenship*. In it he writes that "either no member of the human race has real rights, or else all have the same; he who votes against the rights of another, whatever his religion, color, or sex, thereby abjures his own." The most interesting part of this essay is when he talks about education for women, stating that "education must be the same for women as for men."

Condorcet argues that both men and women should pursue learning in common, not separately. He states it is absurd to exclude women from training for professions, which should be open to both sexes on a competitive basis. He even says that women should have equal opportunities to teach at all levels; "because of their special aptitudes for certain practical arts, theoretical scientific studies would be of particular value to them."

This is the first time someone has publicly advocated for the education of women. It is certainly a new way of thinking in sharp contrast with the old-fashioned beliefs of most people in France.

Tuesday — July 13, 1790

Tomorrow is the first anniversary of the republic. The entire city is invited to the *Fête de la Fédération* on the fields of the Champ de Mars near the military school. My sisters are preparing the dresses they'll wear to this fes-

Hôtel de Ville, city hall of Paris, center of politics and government.

tival. Everybody is excited, except Mother, who is still upset by the memory of the massacres, that terrible episode that resulted in the killing of so many people during the assault on the Bastille. She will take us, of course, but I know Mother's only justification is to see Queen Marie-Antoinette and the royal family who will preside at the festival. Madeleine and I are going to wear new white dresses and straw hats with red, white, and blue ribbons. Angélique wants to wear her long hair down, adorned with a garland of flowers decorated with flowing tricolor streamers that she designed herself. She is so proud of her creation and can't wait until tomorrow to show it off.

Many people from the provinces have arrived in Paris already. The cafés are crowded, the streets are full of visitors; the city is bustling with anticipation. For three weeks many people have worked to prepare the Champ de Mars for the festival. Millie said that women and even children were helping the workers.

Father and his colleagues gathered to discuss the events of yesterday at the meeting of the National Assembly. The delegates passed laws purporting to make all members of the clergy subjects of the State. What this means is that, from now on, the priests will be paid by the government and will be elected by the State or district electors. The priests will have to swear an oath of allegiance to the nation, and the government will regulate their conduct. Any priest who refuses to swear the oath will be forbidden to perform any of his duties and will be subject to arrest and punishment if he does. The priests who swear the required oath will be called government clergy, and those who refuse will be called refractory clergy.

Mother is troubled by this turn of events. She argued that the pope is the head of the Catholic Church, not the Assembly, and that the government should not rule the clergy. Father simply noted that the new Civil Constitution of the Clergy is official and nothing can be done about it. He tried to remind her that, since February, monastic vows were forbidden, so the clergy must be elected by the State, or there won't be priests at all.

The church as we knew it no longer exists. My mother insists that if the priests take the oath, she will not take communion from them. Then, what would she do?

Wednesday — July 14, 1790

We celebrated the Fête de la Fédération in the rain. We awoke early at the sound of thunder. However, not even rain could dampen my sister's enthusiasm for the festivities. After morning prayers, Angélique ran around impatiently, urging everybody to hurry and finish breakfast. She did not want to miss the parade that preceded the festival. Angélique hardly ate, prodding Millie to help her get dressed.

At midmorning, my sister called us excitedly when the first sounds of the military drums were heard in the distance. The parade started at the Temple, advancing along the rue Saint-Denis. Marching to the sound of artillery salvos and military bands were soldiers, the National Guard, government delegates, sailors, and even a battalion of children bearing a banner declaring they were "the hope of the *patrie*." People in the streets dropped flowers as they passed by. My sister and Millie threw kisses and waved at the soldiers who marched proudly wearing wet uniforms and squelching boots.

After the procession turned westward onto the rue Saint-Honoré, we followed it by carriage. We dropped Papa at the Place Louis XV, the plaza where he reunited with the deputies of the National Assembly to join the procession. We kept riding the carriage on our way to the Champ de Mars. It was already one in the afternoon when we arrived, entering the amphitheater through the triumphal arch full of people. A huge crowd—people from every class and all stations of society, workers, bourgeois families, peasants, and aristocrats—waited for the celebration to begin. Music was playing, making everybody merry and gay; many were singing patriotic songs and dancing in the rain.

When King Louis and his court arrived, he sat on a blue velvet throne on a high platform. At his side was Queen Marie-Antoinette, wearing a beautiful red, white, and blue cape with feathers. We were seated not too far from the platform so we could see their faces clearly. There was also a handsome man on horseback, smartly dressed with military insignias. Mother pointed to the man on the white horse to tell us he was the Marquis de La Fayette, the commander of the *Garde Nationale*.

At three thirty, hundreds of priests wearing tricolor sashes over their white albs made an entrance. A *Te Deum* was sung at the special Mass, a hymn written by Joseph Gossec to celebrate national unity. Gossec set a traditional Latin text to music scored for wind instruments, the sound of which made people tremble with emotion. At the end, the bishop of Autun, Charles Talleyrand, blessed the crowd. The Marquis de La Fayette advanced, pointed his sword to the burning flame, and pronounced an oath of fidelity to the nation. All the delegates repeated the oath.

The queen lifted the Dauphin on her arms and people cheered. They responded to the sight of the beautiful child with a demonstration of love for the royal family. Everybody rejoiced in a grand display of unity and brotherhood, waving flags and colorful banners. After the crowds quieted down, the king rose on the platform and pledged to see the laws upheld, and to support the new Constitution of France. After hearing these words, the crowd shouted enthusiastically *"Vive le roi, vive la reine, vive le dauphin!"* My mother was so happy. When I looked up I saw tears forming in her eyes.

On this day the people of France—rich and poor, young and old—were in harmony with the monarchy, with the government, and with one another. I hope this marks the end of the chaos and the social conflicts of the past year.

Sunday — July 18, 1790

The *Fête* continues. This afternoon we saw a splendid water festival on the Seine River, with music, jousting, and dancing. It is now almost midnight, and the party on the riverbanks is still going on. I can hear the sounds of music and the laughter of people. Angélique, Millie, and I sat by the loft windows this evening and watched the fireworks. The silly girls squealed with delight at every whooshing rocket shooting up, and covered their ears at the powerful blasts. The final bouquet of glittering exploding powder was the most spectacular, as it lit the sky with many breathtaking colors.

My parents and sisters have gone to bed. I, however, feel restless and remain in my reverie. It is not that I am disturbed by the clamour of the crowd celebrating near the river. The truth is, when I concentrate on my studies, I lose sight of my surroundings. But tonight I feel agitated, anxious to write down a revelation that came to me unexpectedly.

Some time ago I came across a mathematical relationship that intrigued me, but I did not know its significance until now. When I calculated the compound interest for my father, I used the relationship $A = P(1+R/n)^{nt}$. Then I concluded that, as the value of n increases, the term $(1 + R/n)^n$ is bounded.

It just occurred to me that the expression indeed has a limit and the limit is e, the number discovered by the great mathematician Leonhard Euler. Looking back at my notes, I wrote that e is the limit of $(1 + 1/n)^n$. This expression is exactly like the term in the formula for the compound interest, with $R = 1$. Now I understand why the total amount paid (A) did not increase much. I must tell Papa that there is a well defined mathematical limit to the computation, so he might as well settle for a reasonable number of times a year to compound the interest on his loans.

Très bien. So, if e is the limit of $(1+1/n)^n$, then I can write, for very large value of n (n approaching infinity, i.e., $n \to \infty$), $A = Pe^t$. This expression means that A grows exponentially, and that, even if interest is compounded continuously, the return on a loan is *finite* since the limit of the compound function is e. What a neat way to represent continuous compounding.

I once wondered what e means. Knowing that π represents the ratio of the circumference to the diameter of the circle, what about e? Does it relate to something tangible? At least for the problem of compounding interest, I'd say that the number e represents the limit to greed! Seriously, e seems to represent the limit to growth, which makes sense since otherwise things would expand and grow continuously, without bound, to infinity!

Sunday — September 5, 1790

I waited until now to write, held back by a promise I made to my mother. When we arrived in Lisieux, Mother took away my pencils to discourage my analytical work. At first I resented her, but then I realized that probably it was good idea to let my mind rest for a while. However, during our holiday I read philosophy and history books. In the evenings I read poems and novels to Mother. She was bedridden with some strange ailment for a couple of weeks. Papa stayed behind in Paris, but a week later he came to join us. Papa and I played chess and had a few conversations about philosophy and science.

Angélique and I spent many splendid afternoons watching the peasants work the lands. We saw young girls milking the cows, assisting their mothers making cider, and working in many other farming chores. The women in the village invited us once to help them wrap the cheese they make for the market. Oh, but the best time we had was playing with kites. I taught Angélique how to make diamond-shaped kites using light frames of wood covered with colorful tissue paper. We spent hours flying the kites, watching the streamers undulating vigorously by the strong currents. We ran across the fields to fly the kites higher, laughing when the kites fell down and holding our breath when they flew so far away that we lost them to the wind.

Many nights I spent in solitude, watching the astonishing night sky studded with millions of stars sparkling like diamonds. I wondered about the moon growing and disappearing from the sky. I tracked its progress from a huge luminous disk on July 26 to a thin silver crescent on August 8. The cycle repeated and once again the moon was full two weeks later. We left Lisieux on the first of September, and that night I saw the moon waning.

I am glad for the holiday in the country, and to have the break to witness the splendor of the heavens. Now I am ready to resume my studies, anxious to seek new mathematical challenges.

Friday — September 10, 1790

I began to study infinite series, first trying to understand how they are expanded, such as this:

$$\frac{1}{(x+1)^2} = 1 - 2x + 3x^2 - 4x^3 + \cdots .$$

When I substitute $x = -1$, the left-hand side yields $\frac{1}{(x+1)^2} = \infty$ (since one over zero is infinite). So, for $x = -1$,

$$\infty = 1 + 2 + 3 + 4 + \cdots .$$

This result makes sense; it implies that adding all the integers in the universe would result in an infinite value.

But there is another series that leads to a startling result:

$$\frac{1}{1-x} = 1 + x + x^2 + x^3 + \cdots .$$

When I substitute $x = 2$, the left-hand side of the series is equal to -1, but the right-hand side adds up to a number that would be much larger. That is, for $x = 2$ the series yields,

$$-1 = 1 + 2 + 4 + 8 + \cdots.$$

If I compare the two expressions, the sum of the second series (for -1) seems to be term by term greater than the sum of the first series (for ∞); in other words,

$$-1 > \infty.$$

How is this possible? How can minus one be greater than infinity? But clearly $\infty > 1$. Now, if $x = -1$, the left side of the second series yields $\frac{1}{1-x} = \frac{1}{2}$. Therefore,

$$\frac{1}{2} = 1 - 1 + 1 - 1 + \cdots.$$

This does not seem correct either.

Euler referred to such a seemingly absurd result as "paradoxical." A paradox, according to the dictionary, is "a statement that seems contradictory, unbelievable, or absurd but that may be true in fact." So, perhaps my two contradictory results are mathematically correct; therefore, I should call them mathematical paradoxes.

However, I also wonder whether one should restrict the value of x so that a given series is always true.

Wednesday — September 15, 1790

Mathematical paradoxes have intrigued philosophers since ancient times. Allegedly, the poet Epimenides once said back in the sixth century: "All Cretans are liars." The statement is called the Cretan Paradox because it was uttered by a Cretan, thus is true if and only if it is false. This is the earliest known attempt at formulating a mathematical paradox.

However, it seems that the statement of Epimenides, "All Cretans are liars," is not paradoxical but merely contingent. For example, suppose there are 100 Cretans in the world, one of them being Epimenides. By saying "All Cretans are liars," Epimenides implies that all 100 Cretans are liars; in other words, that all 100 Cretans that exist in the world always make false statements. Could Epimenides be a liar? Yes, simply if one of the other Cretans doesn't always lie. Then Epimendes' statement is simply a false statement, as one would expect from a liar.

Epimenides can't be a truth-teller, because Cretans can only say the truth. "All Cretans are liars" implies "Epimenides is a liar," so if "All Cretans are liars" is true, then so is "Epimenides is a liar," which is a contradiction; hence Epimenides cannot be a truth-teller.

However, if Epimenides was the only Cretan, then "All Cretans are liars" is a paradox, since the statement implies that he is not telling the truth, or that Epimenides is saying, "I am a liar." It is a little confusing; that is why paradoxes are studied in logic.

Now, returning to infinite series, I think I understand why my result seemed paradoxical. Euler wrote about it: "Notable enough, however, are the controversies over the series $1 - 1 + 1 - 1 + 1 - \cdots$ whose sum was given by Leibniz as $\frac{1}{2}$, although others disagree. ...Understanding of this question is to be sought in the word 'sum;' this idea, if thus conceived—namely, the sum of a series is said to be that quantity to which it is brought closer as more terms of the series are taken—has relevance only for convergent series, and we should in general give up the idea of sum for divergent series."

Euler was referring to Gottfried Wilhelm Leibniz, a German philosopher, mathematician, and logician who died when Euler was nine years old. Thus, if Euler and other mathematicians deal with this sort of paradox, I should also understand how to separate a paradox from an untrue mathematical statement. I must be careful and learn to evaluate the results from analysis that may appear contradictory.

Monday — September 20, 1790

Prime numbers are fascinating! Of all the numbers in mathematics, primes are my favorite. Of course I like all numbers, but the primes are more interesting. I am not the only one attracted to prime numbers. Many ancient mathematicians considered them mystical numbers. Pythagoras and his followers, for example, believed that prime numbers had spiritual properties. They liked primes because they are pure and simple, not like the square root of 2 that cannot be expressed exactly as the ratio of two whole numbers.

There was a Prussian mathematician and historian who devoted a great deal of study to prime numbers. His name was Christian Goldbach. He corresponded with Leonhard Euler describing some of his ideas related to prime numbers. In a letter, Goldbach theorized: "Every even number can be written as a sum of two odd primes." This statement is known as *Goldbach's conjecture* because he inferred his statement from uncertain evidence.

I can verify Goldbach's conjecture for any three even numbers chosen arbitrarily, say: 8, 20, and 42.

$$8 = 3 + 5.$$
$$20 = 13 + 7 = 17 + 3.$$
$$42 = 23 + 19 = 29 + 13 = 31 + 11 = 37 + 5.$$

Eh bien, I've written each even number as the sum of two numbers. Now I need to check if 3, 5, 17, and 37 are primes. Using the definition, I derive the first primes. *A prime number is a positive integer $p > 1$ that has no positive integer divisors other than 1 and p itself.* As I determined before, the first primes are: 2, 3, 5, 7, 11, 13, 17, 19, 23, 29, 31, 37, 41, 43, 47, 53, 59, 61, 67, 71, 73, 79, ...

Yes, I can see that 3, 5, 17, and 37 are primes. Indeed, even numbers can be written as a sum of two odd prime numbers, confirming Goldbach's conjecture, at least for the even numbers I chose at random. Furthermore, I notice that there is more than one Goldbach pair (sums of two odd primes), as the number gets larger. Is this to be expected? Goldbach's conjecture says only that there is at least one sum, and it does not to say whether there may be more pairs. But it just seems logical to me that there should be more pairs as the number is greater because there are many more primes available to make up the sums.

Goldbach also predicted that "every sufficiently large integer can be written as the sum of three primes." When he said "sufficiently large," Goldbach must have meant an integer greater or equal to 6 because it does not work with smaller integers. For 6 and beyond, I confirm Goldbach's conjecture:

$$6 = 2 + 2 + 2$$
$$7 = 2 + 2 + 3$$
$$8 = 2 + 3 + 3$$

and so on.

It looks quite simple. However, nobody has proved Goldbach's conjecture. I wonder what approach one would need to develop a proof.

Monday — September 27, 1790

This afternoon Angélique and I rode with my parents to the Palais-Royal. While Papa met with his colleagues at the Café de Foy, Maman took us shopping at a fashionable boutique. She bought a beautiful fan adorned with pictures of the royal family, painted in bright lovely colors. Later Angélique sat for a silhouette portrait at the studio next to the café. Maman wanted me to have my silhouette made as well, but I refused. I do not like portraits of any kind. I do not like my looks. I don't even like to see myself in a mirror. My sister made fun of me and asked rather mockingly, "What do you like, Sophie?" I wanted to scream, yet all I could do was blink a tear away. I turned away and said to myself in a whisper, "I like mathematics!".

I've resumed my studies of infinite series. This topic is difficult and full of unexpected results. Just this evening I learned that the number π appears in many equations that are not related to the circle. Euler discovered connections between π and infinite series. Euler derived some amazing infinite series such as these:

$$\sum_{n=1}^{\infty} \frac{1}{(2n-1)^2} = \frac{1}{1^2} + \frac{1}{3^2} + \frac{1}{5^2} + \frac{1}{7^2} + \cdots = \frac{\pi^2}{8},$$

$$\sum_{n=1}^{\infty} \frac{(-1)^{n+1}}{n^2} = \frac{1}{1^2} - \frac{1}{2^2} + \frac{1}{3^2} - \frac{1}{4^2} + \cdots = \frac{\pi^2}{12}.$$

It is astonishing that π appears in the sum of an infinite series! Of course Euler arrived at those results starting with a trigonometric series, which he rewrote in terms of polynomials, factored in a manner only he could see. To me, just by finding that π is related to infinity makes it clear that π is indeed a number whose exact value will never be known, making it a truly magical number.

How did Euler see the connection between infinite series and π? I've studied Euler's books for months, and I still cannot grasp everything in them. There is so much beauty in his analysis, but Euler does not tell me what kind of revelations helped him discover such splendor.

Friday – October 1, 1790

I met a most unusual girl who loves mathematics probably as much as I do. I met her at Madame Geoffroy's party last night. At first I didn't want to go but now I'm glad, as I learned something interesting to include in my studies. But I am getting ahead of myself. First, I want to record what happened last night and how I ended up chatting with a Spanish girl who studies mathematics.

It started with the invitation to attend a musical soirée at the home of Mme Geoffroy. Maman did not give me a choice, so I dragged myself along and rode with her and my sisters. After the concert they left me alone while they socialized. As always, I felt awkward and uncomfortable. Sitting alone in a corner, I tried to stay away from the idle gossip of the women. Then Madame Geoffroy approached me with a girl I had seen earlier. She was unusual with her black hair and dark eyes contrasting with her modest blue gown. She was presented to me as Mlle Maria Josefa Ibarra de la Villa Real.

After Mme Geoffroy left us, the girl told me she is fifteen, and like me, she is fascinated by mathematics. Maria Josefa complained that her family does not appreciate her talents or support her studies. She told me that, since she was five years old, she could perform calculations in her head and could see patterns in numbers. By the time she was ten, Josefa said, she had mastered algebra and geometry, teaching herself from books in her home library, just like me!

Josefa is from a noble family from Aragon in Spain, but after her father died last year, her mother discovered that the family's wealth was almost exhausted. Thus, to save the family's honour and position in society, the mother arranged an engagement for Josefa to marry a rich duke. The poor girl is terrified but feels that she has no choice but to obey. The wedding was to take place in August but, to gain some time, Josefa proposed a trip to Paris to learn the language and customs, since her fiancé is French. I noted that she spoke perfect French, so Josefa admitted this trip was just an excuse to postpone the wedding.

She also confided her desire to meet the leading mathematicians in Paris. Her dream is to learn from them before she goes back to Spain, although her mother, who accompanies her, may not permit it.

Josefa read many of the books I've studied. She asked me a question about geometry while she described an intriguing mathematical challenge she solved; she called it the Delian problem. I did not have time to ask her for details of her solution because Maman came to fetch me; it was time to leave. I remember reading about the duplication of the cube in *l'Histoire des Mathématiques* de Montucla. I wonder if it is the same problem.

It was very nice to meet this girl who's clearly very smart and is probably ahead of me in her studies of mathematics. And like me she struggles because her talent is not recognized by her family. She fears that once she's married, the desires of her heart and her intellect may be drowned by her new responsibilities.

Would my parents also force me to get married? I hope not. Unlike my sister Madeleine, who was anxious to wed, marriage is not a suitable option for me.

Friday — October 8, 1790

Yesterday we saw a group of men at a street corner shouting political slogans and distributing pamphlets. The men, dressed in long pants without stockings, are members of the revolutionary movement. Because they wear the trousers of the workingman, Papa referred to them as *sans-culottes*. By dressing this way the sans-culottes thumb their noses at the aristocrats.

We visited the residence of the LeBlanc family. At the gathering, the conversation centered on the rummors that circulate in Paris. There is growing opposition to the King's ruling and the adults talked about the changes in the new government. M. LeBlanc complained that during the meetings of the Assembly the most aggressive sans-culottes have become a menace, shouting and complaining against the monarchy. They fight with anyone who opposes their views. Many radical sans-culottes get drunk and disturb people in the streets with their speeches against the king. They brandish their sabers and claim that they will cut off the ears of all enemies of the nation, and, at the first sound of the drum, they are ready to fight.

This morning Millie told my mother that a group of sans-culottes was taunting the women waiting in the long lines at the bakery. They made speeches inciting them to march against King Louis, and they promised to kill all the aristocrats and rich abbots.

Papa contends that not all sans-culottes are bad people. The leaders of the revolutionary movement include lawyers who made significant contributions to the writing of the new Constitution. However, he admitted that some of them seek revenge for social inequalities and real or perceived injustices committed in the past by aristocrats. There are always people who lean to the extreme of any ideology. Many sans-culottes are political extremists, and their actions might hinder, rather than help, the ideals of the new French Republic. That's what I think.

Wednesday — October 20, 1790

Monsieur LeBlanc and I had a brief chat about mathematics. He told me Antoine is studying trigonometry because he wants to be an engineer. M.

LeBlanc asked me whether I knew the trigonometric functions. Of course I know trigonometric functions, but his question made me doubt my knowledge.

After dinner, I went to review my notes and found that I had written the definition: "Trigonometry is the study of angles and of the angular relationships of planar and three-dimensional figures." The word "trigonometry" is derived from the Greek *trigônos*, for triangle, and *metron*, for measure. The trigonometric functions, also called the circular functions are the sine, cosine, tangent, etc., written in mathematical equations as $\sin x$, $\cos x$, $\tan x$, and so forth. The trigonometric functions are related to each other. I encountered the trigonometric functions in Euler's book, where I discovered how they are interrelated, not only with themselves, but also with the exponential and the logarithmic functions.

Without realizing it, I began to study trigonometry the moment I learned the Pythagorean Theorem. I used the relation $z^2 = x^2 + y^2$ to calculate the hypotenuse of right triangles. I remember learning how an angle, in a right triangle, could be calculated by either determining its sine or cosine, depending on whether I knew the opposite side, the adjacent side, plus the hypotenuse, or determining the tangent of the angle if I knew the opposite and adjacent sides of the triangle.

After reviewing my notes, I came up with another startling revelation. Trigonometry is also interrelated with imaginary numbers. From Euler I learned that the number e raised to an imaginary power is equal to a number that has two parts, a real and an imaginary part, which happen to be the trigonometric functions sine and cosine:

$$e^{i\alpha} = \cos\alpha + i\sin\alpha$$

where α is any angle.

In the same manner, for a negative power I can write

$$e^{-i\alpha} = \cos(-\alpha) + i\sin(-\alpha) = \cos\alpha - i\sin\alpha$$

so that

$$\begin{aligned}
e^{i\alpha} &= (\cos\alpha + i\sin\alpha)(\cos\alpha - i\sin\alpha)\\
&= \cos^2\alpha - i^2\sin^2\alpha\\
&= \cos^2\alpha + \sin^2\alpha\\
&= 1,
\end{aligned}$$

where I used the fact that $i^2 = -1$. Thus, $e^{-i\alpha} = \frac{1}{e^{i\alpha}}$. *Voilà!*

Friday — November 12, 1790

What do I get if I raise the number e to a negative power? Exponential functions such as e^x, where x is a positive number, represent growth, just as in the compounded interest. I can also write e^{-x}, since, by the rule of exponents, $e^{-x} = 1/e^x$. In this case, e^{-x} represents the opposite of growth, which I call "decay."

Also, the natural logarithm is the inverse of Euler's number e. Since $x = e^y$, taking the natural log of both sides I obtain $\log x = y$. If I have a function $A = Ce^{ky}$, where C and k are any positive constant numbers, then I rewrite the exponential function as $A/C = e^{ky}$, and taking the natural logarithm of each side of this expression I get

$$\log(A/C) = \log(e^{ky}) = ky.$$

Thus, to solve for y, I simply take the natural logarithm of the ratio A/C and divide by the constant k. This is too easy. I should try another problem.

For example, the equation $\log(y-1)+\log(y+1) = x$, where y represents the unknown. Using the rule of logarithms, $\log a + \log b = \log(a \cdot b)$, I can rewrite the equation as:

$$\log\left[(y-1)(y+1)\right] = x.$$

Also, $(y-1)(y+1) = y^2 - 1$, thus

$$\log(y^2 - 1) = x.$$

To solve for y, I recall that a natural logarithm is the inverse of e; thus, the terms on both sides of the equation are raised to the e power as follows:

$$e^{[\log(y^2-1)]} = e^x.$$

And since $e^{\log a} = a$, the expression reduces to

$$y^2 - 1 = e^x$$

or

$$y^2 = 1 + e^x.$$

And from this expression I can solve for y.

Saturday — November 27, 1790

The deputies of the National Assembly are terribly angry at the dissenting bishops and clergymen who oppose their authority. The Assembly issued the Civil Constitution of the Clergy that requires that every priest swear an oath of loyalty to the State. The oath is meant as a legal means to rid France of the corrupted heads of the Church.

My mother thinks that the oath is a blasphemy because the clergy owe obedience only to the Holy Church. I don't know what to believe. What troubles me is that many citizens are taking out their frustrations on many innocent priests. Some misguided people are breaking into convents and desecrating churches. This is not the way to eradicate the corrupt church leaders. The Assembly should leave alone the churchmen and spiritual leaders who do not oppose the ideals of the new republic.

Saturday — December 25, 1790

It is, at this very moment, terribly cold. And the wind, oh God, the wind! Like an angry ghost howling and pounding on my window. The ink is almost frozen in my pen, making my writing illegible. But I have to record what happened to me this evening. I feel like crying and I wish to disappear into a place where nobody can find me. Why can't I be like my sisters? They are pretty, gregarious, and everybody likes them. I, on the other hand, do not know how to behave and end up making a fool of myself in front of people. We had guests this evening. Madeleine and her husband, Monsieur Lherbette, M. and Mme de Maillard, and M. and Mme LeBlanc and their son Antoine-August came for dinner. I tried to be sophisticated and mature, but I ended up looking like a little girl reciting a foolish tale. Oh, I am so mortified!

It started at dinner. I had not seen Antoine-August for a long time. I could not recognize him, as he's become quite snobbish. At fifteen, he has adopted an air of superiority and speaks so pedantically. He was wearing a lot of *eau de cologne* and powdered curls that made him look like a silly old man. He hardly smiled, trying to appear like an important person. He barely spoke to me throughout dinner.

After the meal we went into the living room. Everybody was in a cheery mood, laughing and having a good time, but I felt awkward and out of place. The party started when Angélique and Madeleine played a lovely sonata.

Maman followed, singing a soulful aria, and then Papa and M. LeBlanc took turns reciting poems from Racine's plays. I was nervous, knowing that my turn would come and someone would ask me to play the piano.

I don't like to play in front of people because I am not as good as my sisters. So, when Maman turned to me, I asked if I could tell them an interesting story. Papa thought it was a great idea, so I chose to tell them about Dido, the queen and founder of Carthage. Antoine, who acts as if he knows everything, asked if this was going to be a long story (he meant boring) and almost made me desist. My father came to me and, holding my hand, invited everybody to listen. I decided to tell the story of the Phoenician princess Dido to avoid playing the piano.

> Dido had to flee from her homeland because her tyrant brother, Pygmalion, had killed her father and was after her wealth. Dido and her loyal followers sailed away until they reached the coast of Africa. Dido wished to find a place to build a city and start all over. When Dido and her people arrived to a place in the coast of Africa, she asked the ruler to sell her a piece of land. For the money Dido offered, the king promised her as much land as might lie within the boundaries of a bull's hide. Any other person would have given up, thinking that the piece of land would be too small, but Dido was a clever woman. She had her slaves cut the hide into many thin strips and tie them together to form a very long one. She then ordered them to lay the long strip over the land, except over the seashore facing the Mediterranean Sea. Dido wanted to secure as extensive a territory as possible within this boundary, so her slaves laid the skin strip into a half-circle over the land. Choosing this shape, Dido got a much bigger piece of land than the king had thought possible. On this land Dido founded the great city of Carthage where she ruled for many years.

There was silence after I finished talking. I am not sure if they understood the significance of Dido's story. I asked them, "How did Dido know that in order to get more territory next to the seashore she had to choose a piece of land in the form of a semicircle?" Yawning, Antoine-August replied that everybody knows that the circle encloses the maximum area. To that my father replied, "But do you know why?" When there was no answer, Papa said proudly, "Well, my Sophie, *ma petite élève*, she certainly knows!"

Antoine-August wanted to show me that I was wrong by half. So, before he did, I explained why Dido placed her strips of bullhide in a semicircle and not in a circle. Since she was using the shoreline as a boundary for her land, she did not need to waste the strip for that part of the boundary. Under that condition, the semicircle gave her indeed the largest area. Of course, if the clever princess would have to choose a piece of land away from the seacoast, she would have selected it in the shape of a circle.

After that uncomfortable moment the party became unbearable for me. No one made any comments but after an awkward silence, everybody resumed their singing. I felt uneasy but nobody noticed; the party continued without me. And even after the guests left I couldn't stop feeling like a fool.

The church bells tolled the midnight hour long ago, and I am still awake, remembering the incident. But after pouring my soul in this writing, I do not feel like crying anymore.

Friday — December 31, 1790

It's past midnight, and Paris is bursting with joyful celebrations. I hear the laughter, the singing from merry folks on their way to balls and dances. Some people amble through dark alleys, seeking the cheery lighted avenues; others ride in their cozy carriages. Today people rejoice; they do not care about affairs of state or economic troubles. On this celebrating night there are no political speeches, no calls to arms.

King Louis-Auguste XVI succumbed to the pressure of the National Assembly and granted his public assent to the Civil Constitution of the Clergy. Now the radical reformers have royal sanction to proceed with their plan for submission of the priests. The revolutionary government will impose formal and public oaths of allegiance from the clergy. At dinner, my parents discussed it, saying that the priests have one more day to decide whether to take the oath. Those who refuse to take it will be removed and replaced by more malleable favorites, or the Assembly will create new priests. It's absurd. And yet, this is the new political reality in my country; the new government is taking control of the Church of France.

My parents say that I am too young to understand politics. Maybe that is why I cannot comprehend what is the fundamental purpose of the revolutionary movement in France. As Papa said tonight, one day it will become clear. But for now my mind must be devoted to the pursuit of my studies.

One year ends and another begins. Just like numbers, one after another in a long infinite sequence, time progresses instant by instant. And life follows its course. . . .

Paris, France
1791

3

Introspection

Sunday — January 2, 1791

The bells for Sunday mass tolled as usual this morning. However, today was no ordinary Sunday. The Catholic ritual at Notre-Dame was transformed into a civil ceremony staged by the government. Upon arrival, I knew something different was about to happen. The church was full with army officers dressed in magnificent tricolor uniforms, and instead of sacred music, an orchestra played a patriotic march. This was the day when many priests took an oath to uphold and obey France's new constitution.

The ceremony began when the officers marched to the altar followed by the priests and bishops. At the altar, the clergy pledged to be loyal to the State. Some people cheered when the priests took the oath, but many more booed. Mother didn't express her disappointment until we came back home. It is difficult for her to see the priests forgo their vows to the Holy Church. Papa tried to justify the need for the oath, saying that the Assembly has no intention of tampering with the people's beliefs or Christian ceremonies. The new Constitution is only intended to create a national French Church devoid of corruption. Mother was not appeased. She fears the rituals will be less about God and more about politics.

Thursday — January 6, 1791

The Feast of the Epiphany is one of Angélique's favorite celebrations. She likes it because she receives presents to commemorate the gifts of the Magi. After morning prayers, Madame Morrell set the table with hot chocolate and a special cake. Angélique had to wait until Madeleine and her husband came for supper to get her gift. They gave her a set of charcoal pencils and a drawing pad. I should not get gifts because I am not a little girl anymore, but Madeleine gave me a set of dominoes made of stone and marble that was supposedly torn from the ruins of the Bastille. I don't play dominoes but I thanked M. Lherbette anyway, as probably he chose it. Maman gave Angélique a little mirror encased in white porcelain, and a comb made of amber. Papa gave her a puppet dressed as a court jester.

M. and Mme Geoffroy came later in the evening. The jovial mood of the party turned into a heated discussion when other friends arrived. The women went into Mother's sitting room to gossip, and I took Angélique to my room to read. Later Millie joined us and made us laugh, telling us some funny stories from a play. Millie loves the *Comédie Italienne*. She says that's the best entertainment in Paris. A little silly and sometimes too loud, but Millie is quite smart and would benefit from learning. Right now, however, young women like her are not educated. In fact, it makes me wonder why only young men have the right to education. Would it not make more sense to educate everybody who wants it? I, for instance, would give anything to attend a university or college to learn mathematics. But my mother says those schools are just for men. Why? I don't understand it. I should have a right to education, too.

Thursday — January 20, 1791

It is bitterly cold. The snow keeps falling, piling on the streets like a thick white blanket covering the ground. The city is silent, so quiet as if devoid of people. The only sound I heard this morning was the murmur from my mother praying. She prayed for King Louis who is ill; it is rumored that he has fever and is coughing up blood. Maman says those are symptoms of tuberculosis.

I advance in my studies. The more I learn, the greater my fascination with numbers. It's amazing how many things one can do with numbers. I like algebra in particular; I can spend hours and hours manipulating equations.

Initially I had the impression that geometry involved only calculation of areas and angles, but I changed my mind. I learned that numbers and equations define all geometric shapes. After meeting the Spanish girl, I came to learn how the Greeks discovered the "golden mean," one of the most beautiful concepts of geometry.

The golden mean arose from a rather simple geometrical figure. The Greeks started with a rectangle of sides equal to 1 and 2. This uncomplicated rectangle can be cut in half by drawing a diagonal whose length is, using the Pythagorean theorem, $D^2 = a^2 + b^2$. The diagonal cuts the rectangle in two parts, which in turn are right triangles.

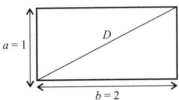

Substituting the numerical value for the sides of the rectangle, I find $D^2 = 1^2 + 2^2 = 5$, so the length of the diagonal is $D = \sqrt{5}$.

The Greek scholars disconnected one of the triangles on one corner of the 1×2 rectangle, straightened out the side equal to 1, and then rotated the side equal to 2 to form two sides of a new rectangle $2(\sqrt{5} + 1)$. By doing so, they discovered that the proportion or ratio between the two sides of the new rectangle is equal to a number with an infinite, non repeating decimal. This I can show as follows

$$\frac{\sqrt{5} + 1}{2} \approx 1.6180339\ldots.$$

Fascinating! This number remainds me of π. Are π and $\frac{\sqrt{5}+1}{2}$ related somehow? If so, how? The two numbers originate from a basic geometrical shape, both are a ratio of two numbers, and both are equal to a number that has infinite, non repeating decimals. Yes, I like to think that the equation connecting these two special numbers must be beautiful and elegant.

That's one of the allures of mathematics. One can discover a myriad of relations of immeasurable beauty. Perhaps I can express my feelings better by quoting the great astronomer and mathematician Johannes Kepler: "Geometry has two great treasures: one is the theorem of Pythagoras; the other, the division of a line into extreme and mean ratio. The first we may compare to a measure of gold; the second we may name a precious jewel."

He was right. How can anybody not find mathematics fascinating?

Saturday — February 5, 1791

Geometry deals with many interesting problems dating back to the mathematicians of antiquity. The problems were challenging for the ancient scholars, not because they did not have solutions, or the solutions they had were unusually hard, but because the known solutions violated an important condition imposed by the Greek mathematicians themselves. For example, the problem known as the "doubling of the cube" requires the construction of a cube whose volume is double that of a given one.

The doubling of the cube is referred to as the Delian problem because it deals with a legend related to the oracle at Delos. According to a letter from Eratosthenes to King Ptolemy of Egypt, Euripides mentioned the Delian problem in one of his tragedies. The story relates that, in 430 B.C., the Athenians consulted the oracle at Delos hoping to stop the plague ravaging their country. Apollo advised them to double the size of his altar, which was designed like a cube. The Athenians tried but could not double the size of the cubic-shaped altar. As a result of several failed attempts to satisfy the god, the pestilence only worsened. Not knowing what to do, the Athenians turned to Plato for advice.

The Athenians could not solve this problem using only a straightedge and a compass, which were the only tools available to solve geometric problems like this one. The Delian problem challenged mathematicians for centuries, due to the restriction of using only those tools. Of course, the solution can be pursued with the help of algebraic analysis.

Eh bien. The doubling of the cube simply means that, starting with the volume of a cube, one must construct a cube whose volume is twice as great. Algebraically, this is reduced to finding a constructible solution to the equation $x^3 = 2a^3$, where a is the length of the cube.

Can I simply show that the equation has no rational solutions? Assuming the opposite is true, let $x = p/q$ be a rational solution. Then $p^3 = 2q^3$. The number of prime factors on the left side of this equation is divisible by 3. The number of prime factors on the right side of the equation, when divided by 3, leaves a remainder of 1. Therefore, the equality is impossible indeed. This means that the equation $x^3 = 2a^3$ has no rational solutions.

With a simple algebraic analysis, the Athenians could have doubled the cube!

Tuesday — February 15, 1791

I can't stop thinking about the Delian problem. The story asserts that the Athenians were advised by Apollo to double the size of his cubic-shaped altar. So, I restate the problem of doubling the cube (the altar) in the form of a binomial $x^3 - 2 = 0$, and try to find a rational solution.

Writing $x^3 = 2a^3$ implies that I have to construct the length x for the cube given by $x = a\sqrt[3]{2}$. If I can find a rational solution, where x and a are integers with no factor in common, I should find that $x^3 = 2a^3$ is an even integer. Since the cube of an even integer is even, x must also be even, ($x = 2p$), which gives $x^3 = 8p^3 = 2a^3$, and $a^3 = 4p^3$. That is, a^3 is an even integer and, consequently, a is also even, and a and x have 2 as a common factor, which contradicts my original hypothesis. Therefore, I can only conclude that the binomial $x^3 - 2 = 0$ is irreducible over the field of rational numbers. In other words, $a\sqrt[3]{2}$ is an irrational number. Just as I had deduced before! Did the Spanish girl arrive at the same solution?

Oh, *mon Dieu*! It's so late. I better get under the covers or Maman will be so angry!

Monday — February 28, 1791

France is in turmoil. Some people oppose King Louis XVI; others support and defend him. A violent confrontation between the two sides erupted today, revolutionaries against royalists, right at the gates of the Tuileries Palace. It was scary. The Marquis de La Fayette, the commander of the National Guard, was away attempting to disperse riotous mobs in the countryside.

At the same time, a group of several hundred aristocrats and supporters of the court, concerned for the safety of the king and armed with guns and knives, gathered and stood guard around the palace. Suspecting that this demonstration was an attempt to help the king escape, the revolutionaries started a violent confrontation. La Fayette had to return in a hurry to disarm the crowd and arrest many of them. Papa was not involved, as he was busy at the meeting of the Assembly. But when we heard the commotion, we were worried for Papa's safety.

Paris is in a state of chaos. But at home we try to preserve a sense of normality. We maintain the same family routines, including the literary evenings and musical interludes that I enjoy so much. I occupy the rest

Angry mob attempts demolition of Vincennes castle — February 28, 1791.

of the time with my studies. And now I have a new activity that brings me much pleasure. I am teaching Millie to read and write. Last night she copied a few paragraphs from a book I gave her for her birthday. She is an eager pupil, so I taught her basic arithmetic. Millie can count and do simple adding and subtracting, but I will teach her to calculate percentages.

Thursday — March 10, 1791

The pope has threatened to excommunicate any of the priests who swear the civil oath. Maman thinks that the decree of His Holiness may stop the madness sweeping France. Swearing an oath to a civil constitution reneges on the most sacred principles. I don't think the message from the pope brings any comfort. On the contrary, his threat puts the priests in a more difficult position, as they have no choice. If they do not abide by the new constitution, the priests will be arrested, imprisoned, and perhaps sentenced to death. Mother is worried but adamant; she will not take communion from any of the government priests because she feels they have betrayed the Church. What a predicament.

I have my mathematics to fulfill me. There is so much to learn! Mathematical analysis requires deep concentration. It also requires that the mathematician observe nature with the eyes of the intellect in order to see how it works. The Greeks knew this. They sought underlying rules and relationships governing numbers and the world around them. The ancient scholars believed that, because the universe was perfect, they could use deductive reasoning to establish truths, without the impurity of empirical measurements. Their approach led to advances in geometry and algebra, and mathematical reasoning became the basis of logical arguments.

I must say, mathematics is the most wonderful and poetic way to discover the universe!

Sunday — March 20, 1791

Angélique draws beautifully. I told her she should be an artist. With her charcoal sketches she captured the beauty of the palaces and buildings in the neighborhood. Now she wants to draw people, and she did a very good job with her first attempt. Millie volunteered to pose for a portrait. I must admit it; the pencil sketch depicts the gentleness of Millie's eyes and her sweet smile. Angélique just had a little problem drawing her nose. Now she is after me, but not meaning to dampen her enthusiasm, I suggested she draw our mother's beautiful face instead and give her the sketch as a birthday present. My sister thought it was a great idea and left me alone.

I am now learning theorems and proofs. What do mathematicians mean when they talk about theorems? Different books say different things, but in simple terms, a theorem is a statement or formula that can be deduced from the axioms of a formal system, by recursive application of its rules of inference. An axiom is a statement that is stipulated to be true for the purpose of constructing a theory in which theorems may be derived by its rules of inference. Well, these are too many words. The best way to illustrate a theorem and its proof is by doing it. First of all, to prove a theorem I need to know certain facts, and then I must follow appropriate mathematical rules.

I start with a very easy mathematical statement: "The sum of any positive number and its reciprocal is greater than or equal to 2." This is a theorem, written in my own simple words. Thus, I must restate it using mathematical notation:

$$\text{If } x > 0, \text{ then } x + \frac{1}{x} \geq 2.$$

This is how I can prove that this statement is true: Since

$$(x - 1)^2 \geq 0.$$

Therefore

$$x^2 - 2x + 1 \geq 0$$

or

$$x^2 + 1 \geq 2x.$$

Hence, dividing by x, I get

$$x + \frac{1}{x} \geq 2.$$

Q.E.D.

Eh bien! This theorem is easy to prove because I know that the square of any real number is nonnegative, and that adding equal quantities to both sides of an equation preserves the inequality.

That's what makes a mathematician. To prove theorems one must know enough mathematics. Also, when mathematicians finish the proof of a theorem they write Q.E.D. as the last statement. When I began to study the book *Introduction to Analysis*, I wondered why Euler always finished his proofs with Q.E.D. Now I know that the three letters are an abbreviation of an expression in Latin, *Quod erat demonstrandum*; it means, "Which was to be demonstrated." From now on, when I finish a proof I will also write Q.E.D.

Friday — March 25, 1791

Mathematics can be challenging. Sometimes mathematicians sense that something is true and may predict an outcome, but they cannot prove it. Learned people can see beyond what is before them and get a notion of what is or could be there. Christian Goldbach, a mathematician who studied prime numbers, became known for his mathematical conjectures. He saw mathematical relations that were not obvious to others. This work is very important because it challenges mathematicians to prove or disprove the statements he made.

I, too, would like to develop such insight and not only make conjectures but also prove them. My satisfaction would be to prove a conjecture or theorem that nobody else has been able to prove. Am I intelligent enough to do it? I am sure I lack the knowledge to do it just now. There is much more to learn.

Sunday — April 3, 1791

The town crier woke us this morning with unexpected news, "Mirabeau is dead!" We ran to the windows to hear the proclamation. The court official went from corner to corner, yelling the distressing news through the neighborhood. Mirabeau, the president of the National Assembly, died last night.

The Comte de Mirabeau was a friend of my father's and came to our home a number of times to discuss politics. Monsieur Mirabeau was one of the supporters of a constitutional monarchy, and he had tried to reconcile the reactionary court of Louis-Auguste XVI with the increasingly radical forces of the revolution. Papa says that Monsieur Mirabeau was in bad health and was under a lot of stress since his election. My mother did not particularly like his way of life, including his gambling. Nonetheless, my mother is sorry to learn of M. Mirabeau's passing. As Papa says, now there is no one in the government as prominent and influential as Mirabeau was, and thus no one who could speak reasonably and so eloquently on behalf of the King.

Thursday — April 7, 1791

I wish I could derive a method to predict primes without having to perform endless arithmetic calculations. But it is not easy. Ever since the discovery of prime numbers, many mathematicians have attempted to formulate a method to generate primes.

In 1640, Pierre de Fermat developed a formula that he thought would produce prime numbers. He stated that numbers of the form $2^n + 1$, where $n = 2^p$ and p is a positive integer, would be prime numbers. However, he was wrong.

It is easy to show that Fermat's formula works for integers 0, 1, 2, 3, and 4, but it fails for $n = 5, 6$, and other values of n. For $n = 32$ (or $p = 5$), the formula gives (according to Euler) $2^{32} + 1 = 4294967297$, a number that can be divided by 641, therefore it is not prime. I do not have the energy or time to calculate the power 2^{32}, so I cannot verify Euler's value. Of course I believe it, but I wonder how he did it, considering that he was partially blind. It is true; he had a phenomenal ability to do complex calculations in his head. Still, this number is too huge to calculate it this way.

In 1644, a friend and adviser of Descartes named Marin Mersenne published a memoir entitled *Cogitata Physica-Mathematica*. In the preface, Mersenne stated that $2^n - 1$ is prime for $n = 2, 3, 5, 7, 13, 17, 19, 31$,

67, 127, and 257, and composite for all other positive integers $n < 257$. Composite numbers are those that can be written as the product of two other numbers.

If I were to assume that all numbers of the form $2^n - 1$ are primes, I would be wrong, because I checked for three arbitrarily chosen values of $n = 2$, 3, and 12 and found that the first two numbers are indeed prime: $2^2 - 1 = 3$ and $2^3 - 1 = 7$. However, it is not true for $n = 12$ because I get $2^{12} - 1 = 4095$. The number 4095 is clearly divisible by 5, so it is not a prime number. This tells me that Mersenne's formula does not generate prime numbers for all values of n.

Oh, I know how arduous it is to calculate such enormous numbers. I wish there was an easy way to uncover prime numbers and wonder, has anyone else developed a different formula that is infallible?

Sunday — April 24, 1791

We went to church early in the morning to celebrate Easter, and in the afternoon we visited M. and Mme Geoffroy. We had dinner with them and a lovely musical soirée. My sisters sang the solemn *Te Deum* to the delight of my parents. A friend of Papa came later with disturbing news. He told us that the royal family was treated very badly.

This morning, King Louis and his family were trying to go to his château in Saint-Cloud. But the National Guard, flanked by a crowd of angry citizens, stopped the royal carriage at the palace gate. The king told the soldiers they just wanted to celebrate Easter Mass in St.-Cloud's chapel, but the soldiers did not allow the carriage to leave. The family wanted to take Easter communion from their own loyal priest. Their reluctance to take the sacrament from a "juror" priest who took the oath to the civil constitution makes the king suspect in the eyes of the government. It is not a good sign.

Saturday — April 30, 1791

Madame de Maillard is one of the most sophisticated and educated women I know. Her knowledge of literature and philosophy is vast. When she speaks, she does it with such passion that one is fascinated listening to her point of view. Madame speaks with an ease, a grace, and an elegance of expression that is captivating. I admire her confidence and her eloquence.

Mme de Maillard believes that women should have the same rights as men, politically and otherwise. She argues that if women had political power, we could change the social system in such a way that it would benefit everybody, not only other women and children. She doesn't think it wrong for me to study mathematics. In fact, she bestows flattering encouragement on me and gives me books or tells me inspiring stories about ancient philosophers. Mother no longer chastises me for studying mathematics; perhaps her tolerance grows by listening to Mme de Maillard, who gives very convincing arguments as to why women should be educated. She reasons that society would benefit greatly if as many women as men would go to the universities, but she admits there is much to be done to change society's notions about the role of women.

I personally do not care for political power. However, without civil rights, women do not have the power to seek and receive the same type of education as men.

Tuesday – May 10, 1791

The Civil Constitution of the Clergy now puts the Church under the control of the State. At tonight's meeting, my father and his friends talked about the new developments. Many priests were arrested because they refused to swear the oath. I just sat silently in my corner with a book before me, although unable to concentrate, thinking about the foolishness of the government's actions. The revolution started with the idea of social equality, so what does that have to do with priests? How is an oath going to uphold those principles? I can understand the appropriation of lands and church treasures, and fighting the corruption among the powerful church leaders. But imprisoning priests for not agreeing to be workers of the government? That is not fair.

After their political discussion, Papa and his colleagues talked about France's scientific position in Europe in the last two centuries. They mentioned the great philosophers and scholars Pascal, Descartes, Mersenne, Desargues, l'Hôpital, and Fermat. One of Papa's friends mentioned the work of contemporary French mathematicians, some of whom are also involved in the political affairs of the new government.

In addition to the Marquis de Condorcet, they mentioned the names Joseph-Louis Lagrange, Pierre-Simon Laplace, Gaspar Monge, and Adrien-Marie Legendre. Papa knows the Marquis de Condorcet very well. In fact, he is the one who sent me the two volumes of Euler's *Introduction to Analysis* last year. M. Condorcet is also known for his activism concerning women's rights after he wrote the essay *On the Admission of Women to the Rights of Citizenship*. He believes that women must share identical educational rights with men. If I remember correctly, many of Papa's friends were against M. Condorcet's ideas regarding women, but in my opinion he is better than most. A man must be quite intelligent in order to accept that a woman is his intellectual equal.

I would like very much to learn the mathematics of the scholars mentioned at the meeting. One day I will muster the courage to write a letter and request permission to attend their public lectures at the Louvre.

Friday — May 20, 1791

Once upon a time, a great man named Don'Raf wandered across a distant land accompanied by his young son. One afternoon, they came upon a stranger sitting by the side of a deserted road. The man was lost, stranded without food, and so Don'Raf offered to help him find his way home. The foreigner accepted and asked for something to eat. Don'Raf responded that he had five loaves of bread and his son had three. Combined, the eight loaves would be shared among the three of them. Thanking them, the stranger promised to recompense them for their generosity with eight golden coins as soon as they reached his home. After eating, the three men mounted the horses and began their journey back to the city.

The next day, upon arrival, Don'Raf and his son were surprised to learn the stranger was in fact the king of that territory. The grateful monarch invited his benefactors to stay the night in his opulent palace. After a splendid supper, he ordered his servant to give the boy three gold coins for the three loaves of bread he shared with him, and five to the father.

But Don'Raf interjected. "Although your division of the coins seems straightforward, it is not mathematically correct. Since I shared five loaves, I should be paid seven coins; my son, who shared three loaves, should get one coin." The king was bewildered, for he did not understand the logic of such argument. Don'Raf offered to demonstrate why his proposal was mathematically correct:

"During the trip, when we were hungry, I took one loaf of bread and divided it into three parts, and each one of us ate a piece. If I contributed five loaves, that means there were fifteen pieces of bread from my bag. And my son contributed three loaves, or nine pieces from his. There was a total of twenty-four pieces of bread, so each one of us ate eight pieces. Now, of the fifteen pieces from my bag, I ate eight, which means I gave away seven pieces. My son had nine pieces and ate eight, so he only gave away one piece of bread. That is why I expect to receive seven coins and my son one."

The king could not argue against such logic; therefore, his attendant distributed the eight coins between father and son according to the father's mathematical formula. Don'Raf thanked the monarch and added: "Your majesty, the division that I proposed, seven coins for me and one for my son, is mathematically correct, but it is not fair in the eyes of God." And gathering the eight coins, the wise man divided them into equal parts: four coins for his son, and four coins he kept.

I composed this tale to entertain my little sister, and also to remind myself that life does not follow an exact mathematical formula.

Saturday — May 28, 1791

Marin Mersenne (1588–1648) was a Minimite friar who contributed greatly to the theory of numbers and other scientific endeavors. In 1611, Mersenne joined the religious order of the Minims to devote time to prayer, study, and scholarship. In 1612, Mersenne became a priest; he taught philosophy at the Minim convent at Nevers from 1614–1618. A year later, he returned to Paris where he lived until his death.

A French theologian, philosopher, and mathematician Mersenne devoted his time to research and published a number of works on acoustics and mathematics. Père Mersenne corresponded extensively with the greatest scholars of his time including Desargues, Fermat, Pascal, Galileo, and Huygens. With Fermat especially Mersenne exchanged ideas in matters related to prime numbers.

Mersenne received many visitors in the monastery here in Paris. He defended Descartes and Galileo against theological criticism, and struggled to expose the pseudosciences of alchemy and astrology. He expanded Galileo's work in acoustics, and with his interest in music, Mersenne provided inspiration and mentoring to the Dutch scholar Christiaan Huygens, the one who wrote *Theory of Music*. Huygens planned to move to Paris in 1646 to be near Mersenne and expand their collaboration. However, the two never met, as Huygens arrived several years after Mersenne had died.

In number theory, Mersenne's work is very important. He tried to find a formula that would represent all primes but, although he failed in this, his work on numbers of the form $2^p - 1$, where p is a prime number, stimulated further developments in number theory.

Friday — June 10, 1791

I feel a little lethargic and physically tired; perhaps it's due to the weather. It is not summer yet but the air feels hot and humid. I spent the day reading biographies. It is fascinating to learn how mathematicians lived, how they studied and developed their mathematics. I started with the *Dictionnaire raisonné des sciences, des arts et des métiers*. My father showed me this encyclopedia when I was a little girl, but I do not remember reading about mathematics at that time. I also found a book about the great scholars of France in the last two centuries.

There was a mathematician who was a child prodigy in mathematics. His name was Guillaume François Antoine de l'Hôpital, marquis de Sainte-Mesme; he was born in Paris in 1661 and died in 1704. When he was fifteen years old, he heard two mathematicians discussing one of Pascal's problems. L'Hôpital astounded them by declaring that he could solve the problem. Indeed, a few days later, he came back with the solution. Being the son of the lieutenant general of the king's armies, l'Hôpital was intended for a military career, but he had to resign because he could not see well. The marquis then devoted himself entirely to mathematics, his favorite subject. When he was thirty years old, l'Hôpital learned mathematics from Jean Bernoulli.

L'Hôpital was a good mathematician and became better known after he wrote in 1696 a book titled *Analyse des infiniment petits pour l'intelligence des lignes courbes*, which was a textbook on differential calculus. This is a branch of mathematics new to me, as it deals with a new concept he calls "*infiniment petits*," which seems to have to do with tiny changes of a function. But I am not sure. In the introduction the Marquis de l'Hôpital acknowledges his mentors Gottfried Wilhelm von Leibniz, Jacques Bernoulli, and Jean Bernoulli who taught him the calculus. L'Hôpital had the greatest teachers!

Saturday — June 18, 1791

The word *brachistochrone* is associated with a mathematical problem that was posed by Jean Bernoulli in 1696. Mathematicians use the brachistochrone to describe the curve between two points, which, for a given type of motion, represents the trajectory that takes the shortest time. An example of this motion is given by the fall of bodies due to the force of gravity. Prior to Bernoulli, Galileo claimed that the natural descent of a body takes the shortest time along the arc that connects the ends of a chord joining two points on a circle. But apparently he was wrong. Bernoulli reintroduced the brachistochrone problem in the scientific journal *Acta eruditorum*.

Bernoulli stated the problem as follows: "Given two points A and B in a vertical plane, what is the curve traced out by a point acted on only by gravity, which starts at A and reaches B in the shortest time?"

Jean Bernoulli knew the solution of the brachistochrone problem, but he challenged others to solve it, too. Mathematicians thrive on challenging not only themselves, but others as well. I've read about this kind of contest throughout the history of mathematics. The challenges are placed to test the mathematicians' methods and the strength of their intellect.

Gottfried Leibniz persuaded Jean Bernoulli to allow a longer time than the six months he had originally intended for solving the brachistochrone problem. Leibniz reasoned that allowing more time would give foreign mathematicians a chance to solve the problem. Several solutions to the brachistochrone problem were submitted by the best mathematicians of the time: Isaac Newton, Jacques Bernoulli, and Leibniz, in addition to Jean Bernoulli himself. They all proved, independently, that the path of least time is an inverted cycloid.

I need to know how they did it. First I had guessed, like Galileo, that the curve traced by the fall of a tiny object would be an arc of a circle (in order to move in the shortest time). Now I know that it is not. There must be something else that I missed, and that is why I need to see what type of equation applies. Maybe Papa will take me to the city library to consult the *Acta eruditorum*. That is the only way to see the solutions to the brachistochrone that the mathematicians submitted.

Monday — June 27, 1791

The king attempted to escape and was caught. This caused a big commotion in Paris. And I witnessed how the royal family was escorted back after their capture. Yesterday, while riding with Mother and Angélique, our carriage had to stop at the Pont-Neuf, as a large group of people blocked the bridge. We did not know what the uproar was about until later in the evening, when Papa came to dinner and told us.

A few days ago Louis-Auguste XVI, Marie-Antoinette, and their children attempted to flee France. The royal family was arrested at Varennes and brought back to Paris. That's what was happening when our carriage was blocked. As we were waiting by the bridge, the royal coach passed us, escorted back to the Tuileries palace. The mob was hysterical! Many people taunted the royal family; men beat on the carriage windows with sticks and pikes.

As the vehicle went by, the men kept their hats on in a gesture of disdain that is so disrespectful. People shouted insults, using such foul language that it made my mother cringe. She kept muttering with tears in her eyes, "Messieurs, in the name of heaven! He is the king of France, for God's sake! Please be kind to her; she is the queen!" As we got closer, we witnessed the rude soldiers pointing their muskets to the ground as the royal family was

Arrival of royal family in Paris after their failed attempt to flee, June 25, 1791.

taken back to the palace. Mother was appalled and deeply troubled by the way people behaved towards the monarchs.

Papa says that after the king attempted to flee France, many people, even those who love and support the monarchy, now think that Louis-Auguste XVI is a traitor and a danger to the nation. The discussion with his colleagues centered on the suspension of the king's authority by the Constituent National Assembly. There is now increased suspicion of royal conspiracies and foreign invasions.

Mother supports King Louis and sympathizes with his attempt to leave a country that no longer respects him. She is troubled to see that the reverence for His Majesty was replaced by contempt, and that the traditional standards that bounded the French people to the king are destroyed. However, one friend of my father clarified that a group of people pasted posters all over, ordering the citizens to keep their hats on when the royal carriage passed by, threatening them with arrest if they didn't. So, it is difficult to say whether the citizens lost their respect for the king, or the leaders of the revolutionary movement are simply manipulating them. Either way the monarchy is crumbling.

Tuesday — July 5, 1791

I attempted to solve a rather challenging problem, but try as I might, I could not find a solution. I was about to give up when, quite by accident, I stumbled upon a book whose author claimed that one cannot be a real mathematician if one does not master Euclid's *Elements*. What a revelation! I had assumed that this ancient book delved in a part of mathematics I thought rather boring because it requires the measurement of geometrical shapes with a ruler and compass. Still, in an effort to overcome the obstacles that prevented me from solving the equation, I decided to study Euclid's geometry.

The *Elements* is a wonderful book for learning algebra and geometry. It is incredible to know it was written centuries ago. Nobody is certain exactly when, but in the introduction there is a note written by the translator saying that, based on a passage in Proclus' *Commentary on the First Book of Euclid's Elements*, Euclid must have lived in Alexandria about three centuries before Christ. That makes the book 2000 years old!

The *Elements* is a set of thirteen books. The first book contains definitions that are fundamental to the study of the mathematics that follows. The first few terms in the *Elements* are primitive terms. Their meanings arise from properties about them assumed known later in the axioms, which are of two kinds: postulates and common notions. Euclid defines a point as "that which has no part." This means that a point has no width, length, or breadth, but has an indivisible location. The first postulate, I.Post.1, for instance, gives more meaning to the term "point." It states that a straight line may be drawn between any two points.

"Line" is the second primitive term in the *Elements*, defined as a set of consecutive points with breadthless length. The description "breadthless length" implies that a line will have one dimension, length, but it does not have breadth or depth. I cannot tell from this definition what kind of line is meant by "line," but later a "straight" line is defined to be a special kind of line. I conclude, then, that "lines" need not be straight. Perhaps "curve" would be a better translation than "line," since for Euclid a curve may or may not be straight. Later, Euclid defines new concepts in terms of others defined earlier.

The subject matter of Book II is called geometric algebra. Since the geometry is plane geometry, the relations represent quadratic equations. This should be easier for me to understand since I know the basics of algebra.

Monday — July 11, 1791

How can a man who died thirteen years ago be now so popular? For the past several weeks people talk about Voltaire as if he were alive. Voltaire was a writer and philosopher, an outspoken critic of the monarchy and of the prelates in the Church. Voltaire's ideas and thinking are now very popular. Many theaters are staging his plays, and his name is invoked whenever someone wishes to justify his opposition to the monarchy.

Because of his criticism of the Church, Voltaire was denied burial in church ground in Paris when he died in 1778. Now, at the urging of the Jacobins—the political leaders who oppose the king—Voltaire's remains were transferred to the *Panthéon*, the mausoleum that was built for the *Grands Hommes* of France.

The coffin was laid at the ruins of the Bastille since last night. And today we watched the funeral procession on rue Saint-Honoré. It was led by a cavalry troop, followed by delegations from schools, clubs, fraternal societies, and groups of actors from the theaters. The workers who demolished the Bastille carried balls and chains from the prison. Four men dressed in classic theater costumes carried a golden statue of Voltaire. Actors waved banners inscribed with the names of Voltaire's plays. A full orchestra preceded the wheeled sarcophagus, which was drawn by twelve white horses. The casket was decorated with theater masks, and the inscription, "Poet, philosopher, historian, he made a great step forward in the human spirit. He prepared us to become free."

Many important people trudged solemnly behind the sarcophagus, including members of the National Assembly, the judiciary and the municipality of Paris. Hundreds of citizens lined the streets to watch the funeral procession, and many people followed it. We did not go to the burial, but Papa said that it was almost midnight when Voltaire's remains were placed to rest at the *Panthéon*.

I am familiar with Voltaire's philosophy, a topic of discussion at home since I was a child. Maman did not like his ideas because Voltaire criticized the king, but Papa agreed with his writing against religious intolerance and persecution. In fact, sometimes he reads Voltaire's plays during our family literary evenings. One of his favorites is *Oedipe*. Papa also enjoys his humorous poetry, an art that Voltaire used to poke fun at the government and nobility.

At one time, Voltaire was imprisoned in the Bastille for writing a mocking satire of the government. After he insulted the powerful nobleman, Chevalier De Rohan, Voltaire was given two options: imprisonment or exile. He chose exile and went to live in England. When he came back to Paris, Voltaire became friends with the Marquise du Châtelet, a very intelligent and erudite lady. They studied natural sciences and wrote books together. Voltaire continued writing books and plays after the death of Madame du Châtelet. He introduced notions of equality and tolerance between men that went beyond social and religious prejudice. Thus, although he is dead, Voltaire's name and his works will live forever.

Friday — July 15, 1791

Mathematicians create mathematical ideas and fascinating worlds independent of the everyday world that we inhabit. Mathematical ideas are figments of the mathematician's imagination, produced by sheer logic and splendid creativity. From Euclid I learned that a point shows only location, but it cannot be seen since it has zero dimensions. Yet, we can see a line segment composed of these invisible points. Euclid also defined a line as infinite in length. The circumference of a circle is also endowed with an ideal perfection, really a curve without thickness and without blemish; but do such figures exist in the realm of our lives? What about a geometric plane, defined as infinite in two dimensions and only one point thick? These mathematical objects can exist only in another world, the world of mathematics.

However, many mathematical ideas are meant to connect to our world, even explain it. An equation can represent something real, like the motion of a body or the change in temperature; that is why I think of mathematics as magic. Is it enough to know mathematics as a pure intellectual recreation? I think not. Knowing how to solve mathematical puzzles is very entertaining, but I want to uncover the secrets of the universe. That is why I must learn other sciences.

If I wish to understand nature, I must know the laws that govern it. Science teaches me the laws; it incorporates basic ideas and theories about how the universe behaves. Science also provides me intellectual pleasure, giving me a framework for learning and pursuing new questions, and gives me an unparalleled view of the magnificent order and symmetry of the universe. Mathematics is the language of science, thus I need to know both.

Wednesday — July 20, 1791

The second anniversary of the destruction of the Bastille was last week. And, tragically, it was marked by a violent confrontation that ended in a horrible massacre, again! The fanfare of last year was missing; there was no royal ceremony, no special outdoors Mass for all the citizens of Paris. This time the sans-culottes held a demonstration on the Champ de Mars to gain signatures for their petition to depose King Louis-Auguste XVI. The demonstration became violent. A contingent of the National Guard (soldiers led by General de La Fayette), fired on the crowd, killing at least fifty demonstrators. This enraged the leaders of the revolutionary movement and fueled their hatred for the monarchy.

Right now the power of the king is suspended until he signs the Constitution. The Assembly is divided. Many deputies support a democratic government with a ruling king. However, the citizens are divided in their support. Right after the Assembly voted to reinstate the powers to the crown, radical groups of sans-culottes began circulating petitions for his dethronement. Papa believes that, instead of promoting peaceful resolutions, the massacre at the Champ de Mars intensified the resolve of the sans-culottes to destroy the monarchy.

As Maman says, political debates can turn friends into enemies. I have seen grown men behave like quarrelsome children, shouting at each other when they disagree about some issue. I should be immune to these verbal matches. I've grown up listening to my father and his friends arguing when they meet every Tuesday night in his library. Lately the discussion is so heated that I no longer want to be around. Some of the men argue in favor of a constitutional monarchy. Others believe that France should become a republic without a king. My father tries to defend Louis XVI, but they all wonder why the king fled, and some claim that this act betrays the people and the nation. Even those who oppose a republic are divided on their support for the monarchy.

Friday — August 5, 1791

Paris is under martial law. There are soldiers everywhere, watching everybody, ready to use their firearms on any citizen attempting to revolt. The soldiers stop people in the streets for no apparent reason, and if suspected of conspiracy, a person can be taken to prison. I feel depressed and confused. There is no rational justification for the bloodshed and the horrors that ravage my country. I spend every day reading the works of great philosophers, trying to understand human behaviour, trying to make sense of the senseless. But nothing justifies the violence and rampant lawlessness.

Paris seems dark and sinister, a city where innocents and guilty alike fear for their lives. After the tragic events of last week I've became painfully aware of the disparity in the revolutionary ideology. France is divided into two groups. The moderates support a democracy without removing the king as the head of the nation. They believe in peace and wish to put an end to the hostility and brutality of the past two years. On the other side are the radical groups who demand the dethronement of King Louis. These radical revolutionaries are ready to use any means, including murder, to achieve their goals. Martial law keeps their anger from boiling over.

My parents try to keep a sense of normality at home, trying not to dwell too much on these awful times. We still have our family gatherings every week. That is the only way to keep us from fretting too much about the events over which we have no control. I am already preparing for tomorrow's get-together. I will talk about ancient Greek philosophers and mathematicians.

In particular, I will discuss Euclid, the most illustrious geometer of antiquity. I wish I could also describe the man. But very little is known about Euclid's life, not even when or where he was born. According to Proclus, Euclid came after the first pupils of Plato and lived during the reign of Ptolemy I. Thus, historians believe that Euclid was educated at Plato's academy in Athens, and that he lived sometime around 300 B.C. Euclid taught at Alexandria in Egypt.

On the personal side, Euclid was probably a kind and patient man. I read an anecdote that reveals a little of his character. The story describes how one day, one of Euclid's pupils after his first geometry lesson asked what he would gain from learning geometry. Euclid told his assistant to give the pupil a coin so he would start gaining from his studies. Very clever, indeed! In another story, King Ptolemy asked Euclid if there was an easier way to learn geometry. At this Euclid replied, "There is no royal road to geometry," and sent the king to study!

When Euclid wrote *Elements*, he integrated all the available knowledge on geometry of his time. Euclid gave a beautiful proof that the number of primes is infinite. He used the methods of exhaustion and *reductio ad absurdum* to prove many theorems. He also discussed the algorithm for finding the greatest common divisor of two numbers.

I could not find anything about the circumstances of Euclid's death. All I know is that he lived and worked in Alexandria for much of his life, and that he established a mathematical school there. It is not much of a biography, but I think this is enough for my presentation about Euclid, one of the greatest and most influential mathematicians in history.

Friday — August 12, 1791

The following is a theorem in the second book of the *Elements*:

> *If a straight line is cut at random, the area of the square on the whole line is equal to the areas of the squares on the segments plus twice the area of the rectangle formed by the segments.*

I translate the statement into an algebraic equation:

$$(a + b)^2 = a^2 + 2ab + b^2.$$

How did Euclid establish such a relationship? His proof is complicated, but I can prove it in a simpler way, if I invoke some rules of algebra:

$$(a + b)^2 = (a + b)(a + b)$$
$$= (a + b)a + (a + b)b$$

$$= a \cdot a + b \cdot a + a \cdot b + b \cdot b$$
$$(a + b)^2 = a^2 + 2ab + b^2.$$

Très bien. Now, how do I generalize the equation to higher exponents, to obtain a formula for the expansion of $(a + b)^n$ for $n = 3, 4, ...$?

The first step should be to obtain an expansion for $(a + b)^3$. I could not find a theorem similar to II.4, giving the expansion of $(a + b)^3$ in any of the Euclid's books. So, following the same method as I did before I get:

$$(a + b)^3 = (a + b)(a + b)^2 = (a + b)(aa + ab + ba + bb)$$
$$= a(aa + ab + ba + bb) + b(aa + ab + ba + bb)$$
$$= aaa + aab + aba + abb + baa + bab + bba + bbb.$$

or

$$(a + b)^3 = a^3 + 3a^2b + 3ab^2 + b^3$$

This is a natural progression that one can follow to arrive at the more general $(a + b)^n$. There must be an easier way to develop the expansion of $(a + b)^n$ rather than the algebraic method I used for $n = 2, 3$. But how?

Friday — August 19, 1791

I have wondered how to prove that there are an infinite number of primes. Euclid thought about that and proved it. He first assumed that there is a finite number of primes, and then proved that statement to be false. The proof is as follows:

Assume there is a finite number of primes $p_1, p_2, p_3, \ldots, p_n$, then $p_1 \cdot p_2 \cdot p_3 \cdots p_n + 1$ is not divisible by any of the p's, so any of its prime divisors yields a new prime number.

Euclid considered only the case $n = 3$. Just like Euclid, I can show that $p_1 \cdot p_2 \cdot p_3 + 1$ is not divisible by any of the p's.

Taking the first three primes: $2 \cdot 3 \cdot 5 + 1 = 31$. Indeed, 31 is not divisible by 2, 3, or 5. However, 31 is a prime number. But I can't do this for all numbers. I will try the proof differently.

Suppose n represents the last prime number. Then I form a new number from the product of all prime numbers up to, and including, the last number n: $2 \cdot 3 \cdot 5 \cdot 7 \cdot 11 \cdots n$.

Now I add 1 to this product and call it k, that is, $k = 2 \cdot 3 \cdot 5 \cdot 7 \cdot 11 \cdots n + 1$. Then k must be prime. If k were not prime, then the list of primes I used to form the product must be missing one. I know that $2, 3, 5, 7, 11, \ldots, n$ cannot divide evenly into k because there would always be one left over every time I tried to divide by any of the numbers $2, 3, 5, 7, 11, \ldots, n$. Therefore, k must be a new prime. This means that the number of primes is infinite, indeed!

Euclid is correct. I could prove his theorem with any number of primes using the same approach. However, that would be too tedious and unnecessary. I'll search for another more rigorous approach to prove that there is an infinite number of prime numbers.

<h1 style="text-align:center">*Tuesday — August 30, 1791*</h1>

I discovered something amazing. I discovered that between the integers 1 and 2 there are at least three irrational numbers: $\sqrt{2}$, ϕ, $\sqrt{3}$. Between 2 and 3 there are at least five irrational numbers: $\sqrt{5}$, $\sqrt{6}$, $\sqrt{7}$, $\sqrt{8}$, and the number e. And between the integers 3 and 4 there are at least seven irrational numbers: π, $\sqrt{10}$, $\sqrt{11}$, $\sqrt{12}$, $\sqrt{13}$, $\sqrt{14}$, and $\sqrt{15}$.

The three numbers ϕ, e, and π, are the most extraordinary irrational numbers I know, each located between two consecutive integers, and the three are found between numbers 1 and 4. Between 4 and 5 one finds the square roots of the integers 16, 17, 18,..., 24. And it makes me wonder if there is another *unique* and *special* irrational located between 4 and 5? If so, it must be a number that relates to something beautiful in the universe. Oh, I am sure of it. Some brilliant mathematician will discover it one day. Or maybe I will find it myself! If I do, I will call it α or ω.

This evening Monsieur de Maillard asked why I study mathematics. I was taken aback by the question and I blurted out, "Mathematics is the most beautiful science!" He looked at me strangely, smiled, and left to join my father. I wanted to explain more, to say that I like mathematics because I enjoy solving equations, and because I take pleasure in pondering the mysteries that lie beneath each theorem. I am intrigued and mystified by numbers and their harmonious connections in equations, like the notes in a beautiful symphony. Yes, mathematics is a rigorous science that demands logic and truth, because it is meant to solve the problems of the universe, including the everyday world around us and the one beyond our reach. I like mathematics because mathematics is truth!

At least I wanted to say that mathematics is a language that speaks to me in beautiful tones. However, I am too shy to express these feelings and thoughts to anyone. I only know that when I study mathematics, I transport myself to another world, a world of exquisite beauty and truth. And in that world I am the person I like to be.

Thursday — September 8, 1791

I woke up at the sound of thunder. Rain is coming, and with it a relief to the hot summer days that refuse to end. The lightning made me nervous so I decided to write to calm me down.

Last night I saw Antoine-August again. Maman gave a dinner party and invited M. and Mme LeBlanc, who showed up with him. I had not seen Antoine in a long time and wanted to know how he was, but I was too shy to ask. Papa, however, inquired about his studies, as if reading my mind. Mme LeBlanc replied that Antoine-August lost interest in mathematics and would rather study law. Well, that's discouraging.

There is much to learn, and there is no one to tell me what topics I must study. However, there is still some hope. Monsieur LeBlanc whispered to my ear that he will insist that his son pursue an engineering degree. Papa and M. LeBlanc commented that Antoine must master mathematics and science in order to pass the rigorous entrance examination. Only the best students are admitted to the *École des Ponts et Chaussées*.

I wish I could be the one preparing for the entrance exams. I would study very hard trying to be the best student in the class. But I know women are not allowed in engineering schools. What a pity. Not that I wish to become an engineer, but I would like to study there or any other school to attend the lectures in science and mathematics given by the best professors in France. For now going to school and being instructed by great mathematicians remain only a dream, a fancy of mine that sustains me through the long days, waiting for a better future.

Wednesday — September 14, 1791

King Louis-Auguste XVI accepted the new Constitution of France. The National Assembly's document provides for an executive—the king—as well as a legislative body. The new Constitution is a moderate document that creates a constitutional monarch and gives the wealthy citizens privileges to a considerable degree.

Papa, who was present at the signing, described a ceremony that took place without the pomp and fanfare that would be expected of such a historical event. He said that King Louis sat in an armchair, not a throne, symbolically placed at the same level as that of the president of the Assembly. Papa remarked that the deputies of the National Assembly kept their hats on during the ceremony, and that Louis XVI was addressed as "sir", instead of "His Majesty."

King Louis, who was raised to believe himself semi-divine, was treated like any other man. Louis-Auguste XVI is no longer the beloved monarch of yesterday. As Papa says, the rapport between the king and the people was broken. Even the soldiers lost respect for the queen. The other day, I spotted a group of them strolling in the Tuileries gardens with their hats on and joking in the presence of Her Majesty. Some soldiers sang disgusting songs as they approached the gates of the palace, as if trying to humiliate Marie-Antoinette on purpose. It is not only insolent, but also hurtful, especially because the soldiers use such vulgar language in front of the children. I am so embarrassed.

Now that the king signed the new Constitution of France, the country may regain some stability. At least today all bitterness was replaced by celebration. There was a spectacular pyrotechnic display this evening, which I saw from my window. I enjoyed the multicolored explosions that lighted the night sky. The citizens of Paris are laughing again, dancing in the streets, full of hope for a better future.

Is the conflict over? Papa claims many royalists declared that they will not adhere to the new establishment because the prisoner King Louis was forced to sign the document under duress. This means the hostility between the groups could escalate.

Sunday — September 18, 1791

We went to Notre-Dame for a special Mass attended by King Louis and Queen Marie-Antoinette. The choir sang *Te Deum* to thank the heavens for inspiring the king to accept the new Constitution of France. Afterwards we went to the Champs-Élysées to partake in the festivities. A hot-air balloon trailing tricolor ribbons glided over the city to announce the good news.

All is quiet now. I am alone in the house, waiting for my parents and my sister, who went to visit Marie-Madeleine. She had a baby boy two days ago. Madeleine and her husband named the baby Jacques-Amant. I stayed behind because I had a terrible headache and needed a nap.

I read a book that describes how, in a letter to Euler, Christian Goldbach conjectured that every odd number is the sum of three primes. When Goldbach wrote that letter, it was still not clear whether the number 1 should be considered a prime, but Goldbach apparently was assuming that it was. In my opinion, if 1 is excluded as a prime number, then Goldbach's conjecture should be rewritten as follows: "Every even number 4 and greater can be expressed as the sum of two primes, and every odd number 7 and greater can be expressed as the sum of three primes."

It looks simple. Yet, after many years mathematicians have not been able to prove Goldbach's conjecture. This is the type of problem that I would like to solve. But I need to know more about prime numbers. Perhaps if I were instructed by a mathematician I could learn faster.

Tuesday — September 20, 1791

For my name day, Madeleine brought me a set of writing quills, and Angélique gave me a beautiful picture she sketched. With her charcoal drawing my sister captured the lively atmosphere of this part of Paris; it shows the Pont-au-Change, the Pont Notre-Dame, and the Quai de la Grève in incredible detail.

Admiring the picture, I realized that Angélique has a keen visual sense, like an artist. In the background she sketched the Conciergerie and the Palais de Justice. She drew the piles of rubble that still remain near the bridge after the houses there were demolished. The most amazing thing is how Angélique caught the movement of people as if the lively scene she saw was frozen in time in her sketch. In the foreground she drew a girl in a white dress with long blonde hair. She also drew a lemonade vendor, a woman selling crêpes, one couple walking their dog, and people walking across the bridges. In her sketch the neighborhood looks charming and gay, just as I like to remember it. On the left corner of the picturesque scene, Angélique drew a group of National Guard soldiers marching with flags waving, heavily armed, a reminder that Paris is now besieged by a revolution. I will treasure my sister's gift forever.

I received more gifts. Millie gave me a lovely white handkerchief embroidered with my initials in red and blue. Madame Morrell fixed a delicious lunch of *pâté en croûte*, mushroom soup, and my favorite chocolate mousse cake.

Mother gave me the best gift of all: she told Papa that I should have a full-size desk in my bedroom. I was speechless. I wanted to hug her and thank her for granting me this wish. I always wanted a large desk, like the one in Papa's library. My lap desk is no longer adequate for writing my extensive notes. I don't know if this is a sign that Maman relented regarding my studies of mathematics. She did not say, and I did not ask her, but I kissed her cheeks tenderly to show my appreciation. Papa was also delighted with the idea, promising to order a desk from the English cabinetmaker, adding that I should have a bookcase to match, to contain my growing book collection.

I am so happy! I will place my desk near the window so that I can have a view of the sky to draw inspiration when I study.

Saturday — September 24, 1791

Mother wanted a chance to get out of the city for some fresh air. She planned a picnic at the *Désert de Retz*, where we had a most enjoyable day. The *Désert* is located about 20 kilometers from Paris. Papa engaged a four-horse berline coach to carry the whole family, including Millie.

The *Désert de Retz* is a gorgeous and unusual garden, which was designed and constructed a few years ago by M. François Racine de Monville, a wealthy aristocrat who is an acquaintance of my father. There is a botanical garden with rare and exotic varieties of trees, ornamental plants, and flowers imported from around the world.

Maman brought a basket of food for our picnic lunch that we ate under the trees. After lunch, Angélique went around looking for inspiration to draw landscape pictures, and I took a long stroll, alone, while my parents conversed. It is wonderful how the mind feels refreshed, walking among the trees, hearing only the sounds of birds singing and the tree leaves moving in the gentle breeze. I imagine that the great thinkers in history developed the most beautiful ideas while ambling in solitude, inspired by the sounds of nature.

Sunday — October 2, 1791

Today is the first day of a new era in the governing of France. The National Assembly dissolved. The deputies accomplished what they had set out to do two years ago. The new Constitution of France was presented to the nation, and the king accepted and signed it. Now that the Assembly's work is complete, a new government was elected.

Monsieur LeBlanc chatted with me yesterday. "Sophie," he said, "imagine I have five books, and I offer to lend you one, how many ways are there of selecting a single book?" The answer was easy, I said, "Five ways." He nodded and continued, "If we label the books A, B, C, D, and E, how many combinations can you make to select two books?" I listed all the book combinations: AB, AC, AD, AE, BC, BD, BE, CD, CE, DE. That makes ten selections. M. LeBlanc agreed with me. He did not give up and asked again, "What about if I lend you three from five?" Well, picking three books from five is the same as discarding two from five so there are ten ways of doing this.

M. LeBlanc was satisfied with my answers. Before leaving to join Papa, he mentioned that Antoine-August is studying this topic, which is called combinations. I stayed behind, trying to understand the relevance of mathematical combinations. I consulted a book that deals with the combination problem. It starts with the binomials. Just as the polynomial is an algebraic expression with multiple terms, a binomial is an algebraic expression with only two terms. The integral powers of the binomial are

$$(x+y)^0 = 1$$
$$(x+y)^1 = 1x + 1y$$
$$(x+y)^2 = 1x^2 + 2xy + 1y^2$$
$$(x+y)^3 = 1x^3 + 3x^2y + 3xy^2 + 1y^3.$$

If I place the coefficients (the numbers on their own and in front of the x's) of the terms in the binomial expressions in a triangular array, with $(x+y)^0$ at the top, the coefficients of $(x+y)^1$ in the second row, those of $(x+y)^2$ in the third row, and so on, the resulting arithmetic array is called "Pascal's Triangle." It looks like this:

$n = 0$						1				
$= 1$					1		1			
$= 2$				1		2		1		
$= 3$			1		3		3		1	
$= 4$		1		4		6		4		1
$= 5$	1		5		10		10		5		1

Pascal's Triangle is named after Blaise Pascal, who described it in his posthumously published book *Traité du triangle arithmétique*. It works for all positive whole number values of the power index. I can show, for example, that,

$$(1+x)^4 = 1 + 4x + 6x^2 + 4x^3 + x^4.$$

Indeed, the coefficients of the binomial expansion with $n = 4$ are the 1, 4, 6, 4, 1 from Pascal's triangle. The triangle has many properties and contains many interesting patterns of numbers. Some simple patterns I see by examining the diagonals in the triangle:

(a) The diagonals going along the left and right edges contain only 1's.

(b) The diagonals next to the edge diagonals contain the natural numbers in order.

(c) Moving inwards, the next pair of diagonals contain the triangular numbers in order.

What if I colored the odd numbers on the triangle, would I get an interesting shape? Pascal used the triangle to solve problems in probability theory. What else could I do with it?

Friday — October 7, 1791

I will expand the following binomial: $(2 + 3x)^5$. Applying the general rule of ascending and descending indexes, and the coefficients from Pascal's Triangle, I can expand as follows:

$$(2+3x)^5 = 2^5 + 5(2^4)(3x) + 10(2^3)(3x)^2 + 10(2^2)(3x)^3 + 5(2)(3x)^4 + (3x)^5.$$

This simplifies to

$$(2 + 3x)^5 = 32 + 240x + 720x^2 + 1080x^3 + 810x^4 + 243x^5.$$

Algebraic expansions of binomial equations are related to the selection of combinations. So, if somebody asks me, "How many ways are there of selecting six objects from eight?" I use Pascal's Triangle and, since there are eight objects, I look at the line beginning with 1, 8, etc. To select six I count along from zero until I get to six. The number there is 28, so there are 28 selections.

Pascal's Triangle is very useful for finding the coefficients in the binomial expansion, especially if the power is $n < 6$. But it gets cumbersome when the exponent is greater, as one page is not big enough to contain all the rows of the triangle. For $n > 7$, it would be better to use a formula to calculate the number of combinations.

I am going to use Pascal's Triangle to play with Angélique. Maybe I can invent a game of combination of cards. This would be an interesting pastime to play with her during the quiet days of winter, when it becomes so cold that we stay indoors for weeks.

Saturday — October 15, 1791

Autumn is beautiful. During these months before winter the light of day seems a little softer, and the sunsets are more colorful. There is also some melancholic feeling in the air, or maybe it is my own nostalgia, realizing that I'm leaving my childhood behind.

My little sister is also growing up. I was so absorbed in my studies that I did not notice the change in her. She traded her childish patter for a more contained countenance. She now moves with grace, without running wildly through the halls like a playful puppy. This evening she came to my bedroom and asked me, "Sophie, what is love?" I was taken aback by her question, for I had never thought about such matters. Angélique seemed flushed when she asked, and was eager for an answer. I did not know what to say, except to hold her hand and say that I love her.

But Angélique desired to know about love between two people who are not related. What *is* indeed that kind of love? I ask: Is love the emotion that my parents felt when they met, the passion that drew my sister Madeleine to her husband? If one pays attention, one sees in people certain tenderness when they are with a dear beloved friend. That is the kind of love Angélique was curious to know about.

After my sister left, I wondered more about her question, but even now I still do not have an answer. I know only mathematics. I could tell Angélique about *nombres amicaux*, amicable numbers. Greek mathematicians defined two whole numbers to be "amicable" if each was the sum of the proper divisors of the other. Reminiscent of a strong friendship between two people, a relationship that defines one person in terms of the other. The best way to illustrate amicable numbers is to give an example.

The pair 220 and 284 has the curious property that each number contains the other, in the sense that the sum of the proper positive divisors of each number adds up to the other. Let's see: the divisors of 220 are 1, 2, 4, 5, 10, 11, 20, 22, 44, 55, 110, and of course, 220 itself; but the last number is self evident, so I take only the other eleven divisors and add them up:

$$1 + 2 + 4 + 5 + 10 + 11 + 20 + 22 + 44 + 55 + 110 = 284.$$

The divisors of 284 are 1, 2, 4, 71, 142, and when I add them up I get 220, that is:

$$1 + 2 + 4 + 71 + 142 = 220.$$

That's what I mean saying 220 and 284 are amicable, as each number "contains" the other.

Amicable numbers were used in magic and astrology and in the making of love potions and talismans. The Pythagoreans knew of them. They called 284 and 220 amicable numbers, adopting virtues and social qualities because the parts of each number have the power to generate the other. It is said that Pythagoras once exclaimed that "a friend is one who is the other I, such as are 220 and 284."

Are there other amicable numbers? Yes! The numbers 17,296 and 18,416 were the next pair of amicables discovered by an Arabic mathematician named Ibn al-Banna in the thirteenth century. Then, in 1638, the French mathematician René Descartes found the more astonishing pair 9,363,584 and 9,437,056. I wonder how the mathematicians found amicable numbers. I haven't found a formula or method to uncover all of the amicable numbers. However, formulas for certain special types were discovered throughout the years.

An ancient mathematician named Thabit ibn Kurrah noted that if $n > 1$ and each of $p = 3 \cdot 2^{n-1} - 1$, $q = 3 \cdot 2^n - 1$, and $r = 9 \cdot 2^{2n-1} - 1$ are prime, then $2^n pq$ and $2^n r$ are amicable numbers. In 1636, Fermat uncovered al-Banna's pair 17,296 and 18,416 using the formula with $n = 4$, as he reported in a letter to Mersenne. Two years later, Descartes wrote to Mersenne with the pair 9,363,584 and 9,437,056 for $n = 7$. By 1747, Euler had found thirty additional pairs, and then he added about thirty more new amicable pairs.

It's impressive. However, none of the books that mention these incredible accomplishments describe how the mathematicians discovered the amicable pairs. Surely, they did not take numbers and check their divisors one by one. This would not be mathematical analysis. There must be a method or an equation that generates amicable numbers.

Oh, but I digress. I started with a question of love and ended up with mathematics. I do not expect that Angélique would understand it.

Monday — November 14, 1791

I am still upset and exasperated. It's hard to put it in words, but I must try to express how insulted I feel by the words of a foolish man. I went with my parents on Saturday to the Théâtre de la Nation to see *La Partie de Chasse de Henri IV*, a comedy in three acts. I was enjoying the play, but an unfortunate encounter at the intermission ruined it all.

There I was with Maman and Papa, taking a refreshment, when M. and

Mme Salvaterry approached us. Mme Salvaterry is a frivolous woman. She kept looking at her reflection in the hall's mirror, arranging her curls while talking nonsense. Her husband is not very nice either. I do not even recall how the conversation moved to my interest in mathematics; all I remember is that M. Salvaterry remarked, without even looking at me, "Ah Monsieur Germain, your clever daughter must read those marvelous books for young ladies. They are simple enough, so maybe she can understand such abstract matters."

I was furious. How dare he suggest I read such inane books? And then his wife added with a forced smile, "They are wonderful, my dear!" and went back to rearranging her hair as if that was all she knew how to do. My father was about to say something but was interrupted by the call for the second act, and we returned to our seats. I am upset that anybody would suggest I read such nonsense.

That night I lay in bed, listening to the wind outside my window, unable to sleep, asking why anyone would think that I'm stupid enough to read those silly books. He was referring to the "Science Books for the Use of Ladies," that a popular author writes believing that women are interested only in romance. So he attempts to explain science and mathematics through the flirtatious dialogue between two silly people in love. Those books are girlish, total nonsense! Oh Papa, why did you not say that I am studying to become a true scholar? I only read books written by real mathematicians.

I wish I could articulate my thoughts and speak my mind. I would have said that I am not interested in frivolities. Alas, I do not know how to say these things without sounding ill-mannered.

Tuesday — November 29, 1791

I have amused myself with binomial expansions, fascinated by their symmetry and how easy it is to find the coefficients. Papa told me that binomials are used in gambling games, and then he told me a little bit about how this part of mathematics was developed.

To begin with, although the arithmetic triangle is named after the French mathematician Blaise Pascal, several other mathematicians knew about and applied their knowledge of the triangle centuries before Pascal was even born. The arithmetical triangle was discovered independently by both the Persians and the Chinese during the eleventh century. The Chinese mathematician Chia Hsien used the triangle to extract square and cube roots of

numbers. After Chia Hsien's discovery of the relationship between extract-
ing roots and the binomial coefficients of the triangle, the Chinese continued work on this topic for many years in their effort to solve higher order
equations.

Of course Pascal developed many of the triangle's properties and applications, making use of the already known binomial coefficients. His work
stemmed from the popularity of gambling. A French nobleman had asked
him what the probability of certain outcomes would be when he threw dice.
Pascal wrote to Fermat about it, and after some discussions and reasoning,
the idea of the arithmetical triangle was born.

Using Pascal's Triangle, one can find the number of ways of choosing k
items from a set of n items simply by looking at the kth entry on the nth row
of the triangle. So, to see how many different trios one could form using the
45 members of a group, one would look at the third entry on the 45th row.
The 1 at the top of the triangle is considered the 0th row, and the first entry
on each row is labeled the 0th entry on the row. The numbers in each row
coincide with the coefficients of the binomial $(1 + x)^n$.

Sunday – December 25, 1791

Christmas is no longer the joyful celebration of my childhood. Of course we
had our traditional dinner and went to church, but there was no *Pastorele*. I
don't know if the new government had anything to do with it, or if the priests
were uneasy to engage in a religious festivity. Maybe the people themselves
are afraid.

Last night I read about Galileo Galilei, a great Italian philosopher and
scholar born in 1564. He prided himself on having been the first to observe
the sky through a telescope. He believed that his greatest quality was his
ability to observe the world at hand to understand the behavior of its components, and to describe them in terms of mathematical proportions. In fact,
Galileo measured the swinging of pendulums until he could describe their
periods by a mathematical law. He rolled bronze balls down inclined planes
to describe the rate of acceleration in free fall. He performed many experiments, and he deduced the mathematical equations that govern the motion
of many things in nature.

In the seventeenth century, the sun and all the planets were thought to
revolve around the Earth. When Galileo learned the new heliocentric theory
proposed by Nicolaus Copernicus, that the Earth and all the other planets

revolve around the sun, he embraced it and became its best proponent. His observations with his new telescope convinced him that indeed the planets revolve around the sun. But accepting the heliocentric theory caused Galileo much trouble with the Roman Church.

In the spring of 1633, Galileo appeared before the Roman Inquisition to be tried on charges of heresy. He was denounced, according to a formal statement, "for holding as true the false doctrine that the sun is the center of the world, and immovable, and that the earth moves!" The great scholar was found guilty and forced to renounce his views. The Inquisition sentenced Galileo to a life of perpetual imprisonment and penance. Due to his advanced age, the Inquisition allowed him to serve the sentence under house arrest at his villa in Arcetri in Florence.

I sit in front of my window, looking at the sky full of brilliant stars shimmering in the darkness of the cold night. The feeble flame of my candle trembles, casting shadows on my writing. My mind wanders. Reading about Galileo confirms my idea of the marvelous intricate connection between mathematics and the universe. I, too, wish to formulate a mathematical law that describes a physical phenomenon. But learning about Galileo's last days also makes me realize that religious beliefs can be in conflict with an intellectual idea. How can a scholar compromise both?

Saturday — December 31, 1791

I wrote an essay to record what happened in 1791, and to reflect on the social changes and my view of the new government. I asked Angélique to draw a picture of Paris to illustrate my writing, to leave a permanent record of what time might erase from our memory. I don't want to forget!

Much happened in the past twelve months. Reading my journal notes, I conclude that 1791 was the year of confusion and violence. King Louis fled Paris, was captured, and returned to Paris. The people were against the king, then supported him, then recognized his power, and then they did not. The king first ignored the new constitution, and then he accepted it, although he did it under duress.

The government also seemed confused. First the deputies supported the king and restored his prerogatives, and then they suspected him of treason. The Constituent Assembly dissolved amidst conflicts and dissensions. Then the Legislative Assembly was formed. Violence erupted throughout 1791, leaving in my memory the horrible image of the massacres of the Champ de Mars, when many people were killed during the protest against the rein-statement of King Louis-Auguste XVI.

It is almost midnight, just a few minutes before a new year begins. I won-der, what does the future hold for the people of France? Will the violence and horrors leave way for reconciliation? If I could have a wish granted, I'd only ask for peace.

Paris, France
1792

4

Under Siege

I begin the new year with more determination and a renewed resolve to study prime numbers. One of my goals is to acquire the necessary mathematical background to prove theorems.

Prime numbers are exquisite. They are whole pure numbers, and I can manipulate them in myriad ways, as pieces on the chessboard. Not all moves are correct but the right ones make you win. Take, for example, the process to uncover primes from whole numbers. Starting with the realization that any whole number n belongs to one of four different categories:

The number is an exact multiple of 4 : $\quad n = 4k$

The number is one more than a multiple of 4 : $\quad n = 4k + 1$

The number is two more than a multiple of 4 : $\quad n = 4k + 2$

The number is three more than a multiple of 4 : $\quad n = 4k + 3$

It is easy to verify that the first and third categories yield only even numbers greater than 4. For example for any number such as $k = 3$, 5, 6, and 7, I write: $n = 4(3) = 12$, and $n = 4(6) = 24$; or $n = 4(5) + 2 = 22$, and $n = 4(7) + 2 = 30$. The resulting numbers clearly are not primes. Thus, I can categorically say that prime numbers cannot be written as $n = 4k$, or $n = 4k + 2$. That leaves the other two categories.

So, a prime number greater than 2 can be written as either $n = 4k + 1$, or $n = 4k + 3$. For example, for $k = 1$ it yields $n = 4(1) + 1 = 5$, and $n = 4(1) + 3 = 7$, both are indeed primes. Does this apply for any k? Can I find primes by using this relation? Take another value such as $k = 11$, so $n = 4(11) + 1 = 45$, and $n = 4(11) + 3 = 47$. Are 45 and 47 prime numbers? Well, I know 47 is a prime number, but 45 is not because it is a whole number that can be written as the product of 9 and 5. So, the relation $n = 4k + 1$ will not produce prime numbers all the time.

Over a hundred years ago Pierre de Fermat concluded that "odd numbers of the form $n = 4k + 3$ cannot be written as a sum of two perfect squares." He asserted simply that $n = 4k + 3 \neq a^2 + b^2$. For example, for $k = 6$, $n = 4(6) + 3 = 27$, and clearly 27 cannot be written as the sum of two perfect squares. I can verify this with any other value of k. But that would not be necessary.

Sunday — January 8, 1792

It's bitterly cold tonight. I hear the wheels of carriages crunching the ice on the streets, sounding as glass breaking. Before going to bed Millie brought me a cup of hot chocolate; she left me a tall candle and restocked my heater. It's after midnight, but I cannot go to sleep just yet.

I found the way to solve a problem that had vexed me for days. It just came to me, so clear, so simple. I know just what to do to prove Fermat's assertion that odd numbers of the form $n = 4k + 3$ cannot be written as a sum of two perfect squares $a^2 + b^2$.

I consider three cases:

Case 1: If both a and b are odd numbers, their squares a^2 and b^2 are odd as well. Therefore, the sum of two odd numbers will be even.

For example $a = 3$ and $b = 5$, so that $a^2 = 9$ and $b^2 = 25$, thus $a^2 + b^2 = 9 + 25 = 34$.

Therefore, I conclude that the odd number $n = 4k + 3$ cannot be written as a sum of perfect squares $a^2 + b^2$ when a and b are odd numbers.

Case 2: If both a and b are even numbers, their squares a^2 and b^2 are even as well so their sum will be even.

For example $a = 2$ and $b = 4$, so that $a^2 = 4$ and $b^2 = 16$, thus $a^2 + b^2 = 4 + 16 = 20$.

Therefore, I conclude that the odd number $n = 4k + 3$ cannot be written as a sum of perfect squares $a^2 + b^2$ when a and b are even numbers.

Case 3: I consider the sum of one square of an even number and one square of an odd number. If a is an even number, it means that it is a multiple of 2, thus $a = 2m$ where m is a whole number. On the other hand, I assume b is an odd number, and as such it is one more than a multiple of 2, or $b = 2r + 1$, where r is a whole number. Therefore,

$$a^2 + b^2 = (2m)^2 + (2r + 1)^2$$
$$= 4m^2 + (4r^2 + 4r + 1)$$
$$= 4(m^2 + r^2 + r) + 1.$$

This means that $a^2 + b^2$ is one more than $4(m^2 + r^2 + r)$. However, it cannot be $n = 4k + 3$ since this is three more than a multiple of 4.

This proves that an odd number of the form $n = 4k + 3$ cannot be written as the sum of two perfect squares $a^2 + b^2$, regardless of whether the two squares are odd or even, or a combination of the two. Q.E.D.

Thursday — January 12, 1792

When I get discouraged, unable to solve a difficult problem, I have to remind myself that other people also had to struggle. It inspires me and helps me not to give up. I've learned that one must be fully devoted to study mathematics, and sometimes one has to persevere against all odds. Many mathematicians probably had the privilege to pursue their studies in great comfort, having tutors to teach and guide them. However, many others had to make a great effort to overcome not only mathematical challenges, but life obstacles as well.

Jean Le Rond d'Alembert, for example, was abandoned by his mother when he was a baby. She left him on the steps of a church in Paris because he was an illegitimate child. The priests found the baby and gave him to a family to raise him. The adopting parents sent the boy to school where his tutors discovered his genius in mathematics. D'Alembert overcame his miserable beginnings and became one of the greatest French mathematicians of this century.

Another great example is Leonhard Euler, a mathematician who had to overcome a physical challenge when he became blind. Yet, losing his sight did not stop his mathematical creativity and genius. I am astonished to learn that it was in the last period of his life, when he was ill and could not see at all that Euler produced some of the most valuable works of mathematics. His friend Daniel Bernoulli had a different challenge, more of a personal nature. Bernoulli loved mathematics since he was a child, but his father ordered him to become a merchant and marry a girl he did not like. Once it became apparent that Daniel was not good in business, his father then forced him to become a physician. Daniel Bernoulli did study medicine, but eventually he dedicated his life to mathematics and he excelled at it.

I can relate to Daniel Bernoulli's struggle. It confirms my belief that if one wants to become a mathematician, nothing can prevent it. Mathematics is an abstraction of the mind, an intellectual desire that cannot be stifled. However, mathematics can't be grasped by a mere glimpse, by simply reading books; it requires serious study and perseverance. I must devote my heart, soul, and every bit of my mental energy to mathematics if I ever wish to master it.

Wednesday — February 1, 1792

Mother is not well. I am worried because her chest pains and headaches are getting more severe. She is always nervous, fretting about Papa; when he is late, she fears he was trampled by the bloodthirsty mobs that roam the streets. Madame de Maillard comes daily to help nurse Mother back to health.

I am fond of Mme de Maillard because she is the only lady who understands my desire to study. She is an intelligent lady, not afraid to express her opinions about the ideals of the revolution. Madame reads the works of great philosophers and can debate anybody with excellent arguments. She says that women are endowed by nature with the same mental capabilities as men. However, Madame also believes that women have a natural duty to remain in the domestic sphere to raise children. She argues that women should be educated in order to rear children intelligently and guide them in their studies. I agree with her about the right to education, especially if she means a right to pursue scientific studies, which is only a privilege for men; but I believe that women should choose whether to raise children or not. I cannot agree on forcing upon women roles they do not wish to play.

I hope I am not catching anything. The chill of winter lingers in my body. My hands are always cold, my legs ache, and I have a cough. During the day I study near the fireplace in my father's library to stay warm. I wear two pairs of wool stockings to bed, and Millie has to warm my blankets with the hot iron. But even if it is cold, I prefer to work at night, when the only sound in the house is the chiming of the clock announcing each hour.

The night is mine to lose myself in the secrets of mathematics. When I master a new topic I feel excited, and my dreams carry me to new dimensions. When I go to sleep without solving a problem or am unable to understand a new concept, dreams do not come easily, just like tonight. It is cold, and my body aches, but my mind is restless, thinking and thinking about numbers.

Friday — February 24, 1792

The snow covers the bruises of the city. Yes, Paris is wounded, tormented by the ravages of social chaos, bloodshed, and economic troubles. The winter is harsh, aggravating the devastation and the misery of the poorest people. It's a difficult time for France now.

Millie complains how hard it is to find groceries. Robberies make it more difficult for the shop owners. Scamps run off with stolen goods and sell them at the other end of the street. Today, Millie caught two thieves loaded with sugar and coffee running along the rue des Lombards. The shoplifters approached her and tried to sell her the sugar at exorbitant prices. Even candles are scarce, and oil for the lamps is very expensive. I don't want to aggravate the situation so I've given up studying at night to save my candles.

My mother is slowly recovering from her illness. Her poor health, the gray bitter winter, and the tumultuous political revolution that goes on in Paris have contributed to the collective depression that surrounds us. Even Millie, usually so vivacious, comes home from her errands not only tired and chilled but also depressed. Without Maman's musical soirées to entertain us every week, the house feels colder and darker than usual. If it wasn't for Angélique's chattering enthusiasm, we'd be miserable.

Saturday — March 3, 1792

It was a lovely party. Some of our friends joined my family this evening to partake in a musical interlude planned especially for Maman. A few weeks ago Angélique came up with the idea of doing something to cheer her up.

I suggested staging the play *Esther* by Racine because it's a favorite of my mother's. This play tells the story of Haman and the Jewish queen Esther who risks her own life to save her people from certain destruction.

Angélique got Millie all excited about performing in the play. She wanted to give me a part as well but I refused, proposing instead to help them in other ways. My sister chose to play Esther and gave Millie the part of King Ahasuerus. With Madame Morrell's assistance, they made the costumes and I helped Angélique and Millie memorize their lines and practice their dialogue. I volunteered to play the musical interludes at the piano because there was no one else to do it.

The play was a welcomed respite for everybody who came to see it. Angélique invited Madeleine, her husband, and M. and Mme de Maillard. Maman did not know about the play, thus she was pleasantly surprised when Mme Morrell opened the drapes to reveal the improvised stage she and her husband helped my sister to decorate. Maman was ecstatic, and I noticed a few tears in her eyes during the monologue when my sister recited the words, "Oh God, my king, behold me trembling and alone before thee!" She was convincing as Esther. Maman was so pleased; at the end of the performance she broke out in applause along with the others.

Tuesday — March 13, 1792

I found a most peculiar book in Papa's library. The first page starts with the following statement:

> "A certain man put a pair of rabbits in a cage. How many pairs of rabbits can be produced from that pair in a year if it is supposed that every month each pair begets a new pair which from the second month on becomes productive?"

It was written hundreds of years ago by a mathematician known as Fibonacci. At first I thought it odd that a book written by a mathematician should deal with such an infantile problem. But soon the puzzle caught my interest, and I decided to learn more about it. Leonardo Pisano, better known by his familiar name Fibonacci, published in 1202 a book titled *Liber Abaci (Book of the Abacus),* based on the arithmetic and algebra that he learned in his trips to far away countries.

Fibonacci introduced the Hindu-Arabic decimal system and the use of Arabic numerals into Europe. His book also included simultaneous linear equations. The second section of *Liber Abaci* contained a large collection of problems aimed at merchants. They relate to the price of goods, describing how to calculate profit on transactions, how to convert between the various currencies in use in Mediterranean countries, and many other practical problems.

The problem of the rabbits appeared in the third section of *Liber Abaci*. The original problem that Fibonacci investigated was, simply stated, how fast rabbits could breed. At first the problem seems inconsequential, but it requires logical thinking to solve it. This is the way I analyze the problem:

Suppose the man starts with a newly born pair of rabbits, one male and one female, and he places them in a cage. Rabbits are able to mate at the age of one month, so at the end of the second month a female can produce another pair of rabbits. Suppose that the rabbits never die and that the female always produces one new pair (one male, one female) every month from the second month on. Fibonacci asked, "How many pairs will there be in one year?" Well, I will try to answer:

1. At the end of the first month, the rabbits mate. There is now 1 pair.

2. At the end of the second month the female produces a new pair, so now there are 2 pairs of rabbits in the cage.

3. At the end of the third month, the original female produces a second pair, making 3 pairs in all in the cage.

4. At the end of the fourth month, the original female has produced yet another new pair, the female born two months ago produces her first pair also, so there are now 5 pairs in the cage.

The number of pairs of rabbits in the cage at the start of each month is 1, 1, 2, 3, 5, 8, 13, 21, 34, 55, 89, 144. These numbers are called Fibonacci's numbers. And the answer to the original question is 144; that is, there will be 144 rabbits in the cage at the end of 12 months.

This problem does not seem very realistic. It implies that sibling rabbits can mate. I would rather say that the female rabbit of each pair mates with any non-brother male and produces another pair. Also, the problem implies that each birth is of exactly two rabbits, one male and one female. This is not true in life. Perhaps Fibonacci used rabbits as an illustration in the sense of mathematical objects rather than as living creatures. In fact, he used fictitious animals to illustrate other problems, such as, "A spider climbs so many feet up a wall each day and slips back a fixed number each night, how many days does the spider take to climb the wall?" And, "A hound whose speed increases arithmetically chases a hare whose speed also increases arithmetically, how far do they travel before the hound catches the hare?"

How important are Fibonacci's numbers in mathematics? Or are they only a game, an intellectual diversion? I don't know, but I can write the numbers in a mathematical sequence and seek a pattern.

To obtain each number of the Fibonacci series, I add the two numbers that came before it. In other words, each number of the series is the sum of the two numbers preceding it. Starting with 0 and 1, I add them to get the next number: $0 + 1 = 1$. Then the next term is $1 + 1 = 2$, and the next is $1 + 2 = 3$, and the series continues. I represent this series Fib(n) as follows:

$$n: \quad 0\ 1\ 2\ 3\ 4\ 5\ 6\ 7\ 8\ 9\ 10\ 11\ 12\ 13\ 14\ 15\ 16 \ldots$$
$$\text{Fib}(n): \quad 0\ 1\ 1\ 2\ 3\ 5\ 8\ 13\ 21\ 34\ 55\ 89\ 144\ 233\ 377\ 610\ 987 \ldots.$$

Trés bien. The sequence of numbers gives a peculiar series that keeps growing, as long as I add the new number and the last. There is a certain rhythm to it: two odd numbers followed by one even; it makes me wonder. What if instead of 0 and 1 I started with any two other numbers, say 1 and 3? Then the series would be $1, 3, 4, 7, 11, 18, 29$, and so on.

Wednesday — March 21, 1792

What a glorious day! Although there is still a chill in the air, the sun came out, and the streets were full of people. Since waking up, I felt euphoric, in a cheery mood. After dinner I had a long conversation with Papa. I asked him if he would take me one day to attend the lectures at the Academy of Sciences. I know that every week mathematicians present their work in the king's hall. It would be the most exciting experience of my life to listen to these great men talk about their scholarly work. I am so curious to hear how they explain mathematics and learn how they apply the analysis to the study of nature. Even though the lectures are public, one needs special tickets, which are reserved for other scholars and important visitors. I'd give anything to gain admittance to the lectures.

I heard again a conversation about a mathematician named Joseph-Louis Lagrange. He is originally from Turin, Italy, but M. Lagrange worked in Germany and, after being recommended by French mathematicians, the king invited him to take a post in the Academy of Sciences here in Paris. Father has an acquaintance who met M. Lagrange and referred to the scholar as a very polite, soft-spoken, and humble person. If I had the opportunity to talk with M. Lagrange, I would ask him many questions about mathematics.

Palm Sunday — April 1, 1792

To celebrate my sixteenth birthday I had a *tête-à-tête* with Papa. We talked about my studies and his hope for a better future. Lately, I've felt so gloomy, and to cheer me up Papa gave me a lovely fan imported from England. But a material gift can't cure my melancholy.

I also miss Angélique's chatter. My sister has influenza and has stayed in bed for over a week. I cannot visit her because Maman is afraid her illness is contagious. I've spent the time alone, reading and translating the last chapters of Euler's *Introduction to Analysis*. It took me a very long time, but I am almost finished.

Mme de Maillard gave me the *Pensées*, a book written by Blaise Pascal, the same French mathematician who studied the arithmetic triangle. The *Pensées* is a collection of hundreds of notes Pascal made, many of them intended for a book that he expected would help him defend Christianity. Pascal organized and classified some notes before he died, but others remain unsorted, and the book he intended was never written. Pascal's *Pensées* are powerfully insightful thoughts that lead us more deeply into contemplation of human nature and the strivings of the heart and mind. He also pondered the visible world, his thoughts tinted for sure by his knowledge of geometry. For example, *"C'est une sphère infinie dont le centre est partout, la circonférence nulle part,"* stating that the world we see is an infinite sphere, the centre of which is everywhere, the circumference nowhere.

Traditionally on this day we start the celebration of Holy Week with a special Mass at Notre-Dame. But under the new government rules the rite was suspended. The churches in Paris are now almost empty; the faithful, like my mother, refuse to take communion from the juror priests, and the rest of the people are angry with the clergy. What would Pascal think? I find solace reading the philosophy of a mathematician who defended the Church.

Easter Sunday — April 8, 1792

Perfect numbers have fascinated mathematicians since ancient times. Many scholars were concerned with the relationship of a number and the sum of its divisors, often giving mystic interpretations. A positive integer n is called a perfect number if it is equal to the sum of all of its positive divisors, excluding n itself.

Perfect numbers have the form $2^{n-1}(2^n - 1)$, where n and $(2^n - 1)$ are prime numbers. To see how this works, I start with the first prime number $n = 2$:

$$2^{2-1}(2^2 - 1) = 2^1(4 - 1) = 2(3) = 6.$$

This shows that 6 is the first perfect number because $6 = 1 + 2 + 3$. The next number is 28 since $1 + 2 + 4 + 7 + 14 = 28$. The next two perfect numbers are 496 and 8128. These four perfect numbers were all known before the time of Christ. More than two thousand years ago, Euclid explained a method for finding perfect numbers that is based on the concept of prime numbers.

There are two theorems related to perfect numbers:

Theorem I: *k is an even perfect number if and only if it has the form $2^{n-1}(2^n - 1)$, and $(2^n - 1)$ is also prime.*

Theorem II: *If $(2^n - 1)$ is prime, then so is n.*

The first perfect numbers 6, 28, 496, and 8128 are even numbers. Are there odd perfect numbers? I could try to verify this by applying the formula $2^{n-1}(2^n - 1)$ to each prime number one by one. However, this is an enormous task, considering that the fourth perfect number is already a large number; it would probably take all my lifetime to verify it. Besides, this is not the way to prove a mathematical proposition. A theorem is true precisely in the sense that it possesses a proof. Thus, to establish a mathematical statement as a theorem, one must demonstrate the existence of a line of reasoning from axioms in the system (and other, already established theorems) to the given statement.

I would like to prove that there are infinitely many perfect numbers. I am sure it would require first proving that there are infinitely many prime numbers. For the time being, I conclude that even perfect numbers have the form $2^{n-1}(2^n - 1)$, where the number $(2^n - 1)$ is itself a prime number, and the exponent n is also prime.

How I wish I could show my analysis to someone who understands the analysis of prime numbers. If I show it to Papa he will be impressed, but he will not know if my approach is correct.

Friday — April 27, 1792

Ancient Greek mathematicians studied prime numbers and their properties. Euclid's *Elements* contains several important results about primes. In Book IX of the *Elements*, Euclid proved that there are infinitely many prime numbers. His proof uses the method of contradiction to establish a result. Euclid showed that if the number $(2^n - 1)$ is prime, then $2^{n-1}(2^n - 1)$ is a perfect number.

Sometime in the year 200 B.C., Eratosthenes devised an algorithm for calculating primes, a method now known as the Sieve of Eratosthenes, so named because it filters out multiples of numbers from the set of natural numbers (excluding 1), so only the prime numbers remain. There is a long gap in the history of prime numbers that lasted several hundred years until 1651, when the French mathematician Pierre de Fermat revived the study of prime numbers. He corresponded with other mathematicians, in particular with Marin Mersenne, who was trying to find a formula that would represent all primes.

Born in Beaumont-de-Lomagne in 1601, Pierre de Fermat was a man of great erudition. He received a degree in civil law, and in 1631 Fermat obtained the post of councilor for the local parliament at Toulouse. He was successful in this position, becoming Royal advisor; however, Fermat devoted all his leisure time to mathematics.

It is truly astonishing that, although he pursued mathematics as an amateur, Fermat's work was of such exceptional quality and scholarship that he is regarded as one of the greatest mathematicians of his time. Fermat was in contact with men of learning everywhere. His correspondence with Blaise Pascal, Christiaan Huygens, and Marin Mersenne contained the results of his mathematical research. Fermat worked in analytical geometry, maxima and minima problems, and in the theory of probability.

The theory of numbers was perhaps the favorite study of Fermat. He prepared an edition of Diophantus, and the notes and comments thereon contain numerous theorems of considerable elegance. Many of his results were found after his death on loose sheets of paper or written in the margins of books which he had read and annotated. In one letter to Mersenne, Fermat conjectured that the numbers $2^n + 1$ were always prime if n is a power of 2. Fermat verified this formula for $n = 1, 2, 4, 8$, and 16.

Mersenne also studied numbers of the form $2^n - 1$, discovering that not all numbers of this form, with n prime, are prime numbers. For example, $2^{11} - 1 = 2047 = (23) \cdot (89)$ is a composite number. For many years numbers of this form provided the largest known primes. How extraordinary!

Tuesday – May 15, 1792

Is there a God? Is there an intelligent Being that created the universe and everything in it? I am consumed with questions of this type but cannot answer them. Sometimes I am not sure what to believe. The problem is that I doubt so much. Ever since I was a child I have prayed, as Mother taught me. But as I grew up I began to question it, thinking it may be futile.

Maman has always said that God answers our prayers. So, when I was a little girl I prayed to God to make me as lovely as my sister Madeleine. But every time I looked in the mirror, as Maman fixed my hair, I did not see any change. I also wondered about the unfortunate beggars sitting at the church steps; many of them are blind, lame, and diseased. Why would God allow such suffering and misery? When Maman was ill I prayed that she'd get better, and when she did I was not sure if she was cured because we prayed, or because she took the tonics the doctor prescribed. Once I even considered becoming a nun, but then I realized that it would be hypocritical since I am not sure of my faith.

My mother and sisters believe in a God who is omnipotent. They seem to find comfort praying to a God that no one sees, but they believe He *is* everywhere. We have witnessed the desolation and the terror that has devastated our country in the last few years. We have prayed for others, whether they are innocent or guilty. We have prayed, but I am not sure it helps.

I read books written by French intellectuals who upheld the principles of Liberalism. Basically they say in their writings that man is responsible to no authority; that men owe nothing to God; and that the mind and will of man replaces the will of God. These three tenets have been discussed many times in Father's library. Ever since I was a little girl, sitting quietly and listening to the adults talking, I tried to understand what it all means. I never see Father pray and wonder whether he, like me, feels any doubt when Mother says that God will save us all. It is hard to believe after we witness the bloodshed and injustices all around us. Yet, I do hope Mother's prayers are answered and that the violence in Paris will end soon.

Wednesday – May 30, 1792

Papa gave me a beautiful book written by Leonhard Euler. It is titled *Lettres à une princesse d'Allemagne sur divers sujets de physique & de philosophie*. Papa explained that Euler wrote in this book the topics he taught to young princess, a girl like me. During his stay in Berlin, Euler was asked to provide tuition for Princess d'Anhalt Dessau, a niece of Frederick the Great. Euler instructed the princess through these letters.

At the start of their correspondence, Euler's royal pupil was a fairly ignorant girl. The princess did not know natural philosophy or mathematics. Euler began to teach her starting with the basic notions of distance, time, and velocity. The letters continued with more difficult aspects of physics: light and color, sound, gravity, electricity, and magnetism. Euler then addressed the nature of matter and the origin of forces. The instruction of the German princess was extensive. The *Lettres* also cover metaphysics, dealing with the mind-body problem, free will and determinism, the nature of spirits, and the operation of Providence in the world of nature. Euler believed that, unaided, human reason will not take us very far, and that some philosophical questions will be left unanswered awaiting divine illumination.

I will study this book diligently. I will pretend that each letter is addressed to me and imagine Euler patiently writing the concepts that he wishes me to learn.

Tuesday — June 5, 1792

Last night we went to the opera. Mother made me wear a new blue silk dress that was most uncomfortable. The full fichu of sheer organza made my neck itch. Maman insisted that I wear a corset too, so I had to sit uneasily upright through the evening. At least I did not have to wear the enormous pannier skirts that women wore years ago. Millie fixed my hair and, against my protests, Angélique put a little rouge on my cheeks. I am glad that they didn't powder my hair and pile it so high that it'd look grotesque. Papa gave me a very pretty fan for my birthday in April, so I took it with me to the concert. Angélique looked all grown up with her white gown and tricolor sash around her waist. She wore her hair curled just like Madame Royale, the young daughter of King Louis.

The opera *Les Evénemens Imprévus* had Mme Dugazon playing the *soubrette*, maid. We sat not far from the royal box, and Mother was very excited, looking forward to seeing the beautiful queen. Marie-Antoinette soon showed up accompanied by the Dauphin and her daughter. Madame Elisabeth, the king's sister, and Madame Tourzelle, the governess, were also with her. I noticed from the beginning that Her Majesty seemed distressed. She was overwhelmed by the applause of her loyal subjects, and I saw her dab the tears from her eyes. The little Dauphin, who sat on the queen's lap, seemed anxious to know the cause of his mother's weeping.

In one act of the opera, Madame Dugazon sings a duet with the valet, and when she sang "*Ah! Comme j'aime ma maîtresse*" she looked directly at Marie-Antoinette, as a sign of affection. At that moment a few angry citizens jumped upon the stage, attempting to hurt Mme Dugazon. The other actors had to restrain the ruffians. During the commotion, the queen and her family left in a hurry. Maman was mortified and upset to see more and more people openly offend the queen. How can they show their scorn so blatantly?

There is a full moon tonight. I will snuff out my candle and sit by my bedroom window to watch the dark sky full of stars. Maybe I will recognize some constellations. On a night like this I feel drawn by the mysteries of the universe.

Sunday — June 10, 1792

The numbers π and e occur everywhere in mathematics. Many mathematicians are fascinated by these numbers and attempt to establish their properties. Some time ago I tried to prove that the number e is irrational, but I could not do it. I needed to learn some mathematical truths in order to make the appropriate assumptions. I try again and begin by stating a simple theorem: *e is irrational.*

Proof. I know that for any integer n

$$e = 1 + \frac{1}{1!} + \frac{1}{2!} + \frac{1}{3!} + \cdots + \frac{1}{n!} + R_n$$

where $0 < R_n < \frac{3}{(n+1)!}$ (just as I saw in a book).

I assume that e is rational, i.e., that there are two positive integers a and b such that $e = a/b$, and I let $n > b$. Then I rewrite the series as

$$\frac{a}{b} = 1 + \frac{1}{1!} + \frac{1}{2!} + \frac{1}{3!} + \cdots + \frac{1}{n!} + R_n$$

and, multiplying both sides by n factorial, or $n!$,

$$n!\frac{a}{b} = n! + \frac{n!}{1!} + \frac{n!}{2!} + \frac{n!}{3!} + \cdots + 1 + n!R_n.$$

Since $n > b$, the left side of this expression represents an integer, therefore

$n! R_n$ is also an integer. But I know that

$$0 < n! R_n < \frac{3n!}{(n+1)!} = \frac{3}{(n+1)!}.$$

So, if n is large enough, the left side of the above expression is less than 1. But then the product $n! R_n$ must be a positive integer less than 1, which is a contradiction. Therefore, e is an irrational number. Q.E.D.

Thursday — June 21, 1792

A gang of angry sans-culottes invaded the Tuileries Palace. It was a brutal attack. Yesterday at around one o'clock, we were coming from visiting Mme LeBlanc when a group of sans-culottes blocked our carriage. Led by Santerre, the armed men were walking in the direction of the royal palace, rowdily shouting insults, and singing the dreaded *Ça Ira*. The awful lyrics incite people to hang the nobility and the clergy. The chanting sounded so menacing that I knew something dreadful was about to happen.

I tried to forget the incident, but later Papa described in vivid detail what happened. The sans-culottes assaulted the Palais-Royal, brandishing pistols and sabers in the king's face, shouting insults, terrorizing the royal family. For several hours, the men humiliated the king and queen and threatened them with violence. They forced King Louis to don the liberty cap, and to drink to the health of the people. The sans-culottes showed no respect and treated the king as if he were a puppet, addressing him as "monsieur" rather than as "His Majesty." The people of France have lost respect for King Louis. Gone are the days when he was treated with the highest reverence. This attack may be the beginning of the end of the reign of Louis XVI.

The tension escalates. There is confusion regarding the law on the deportation of refractory priests, the king's veto of this law, the dismantling of the royal Constitutional Garde, and the stationing of thousands of troops by the Assembly in Paris. Last week the king dismissed his Jacobin ministers and replaced them with more moderate Feuillants. Amidst all this, the violence in the streets escalates. I am afraid the monarchy in France is over.

Saturday — June 30, 1792

It's terribly hot today. I wish we could go on vacation to our country home in Lisieux. I wish we could get away from the turbulence that rolls over

Paris. But now traveling is restricted; people need passports to leave the city, and the situation in the villages is rather unstable.

For the same reason we hardly go to the Tuileries gardens or any other public place, as it is not possible to take a leisurely walk for fear of the mobs. My sister Angélique is getting more miserable sitting in the house all day. I suggested to Papa to take us to an art gallery for my sister's sake. Admiring the works of art will help her with her painting and ease her restlessness. She was thrilled with my suggestion and kissed my cheeks several times.

The only other outing for Angélique is to visit the dress shops in the rue des Petits-Champs with Mother. From time to time there is a little commotion in that part of the city, but it is usually just the shouting of angry shop owners expelling people that linger in front of their windows a little longer than normal, or because the women argue about the cost of gowns they cannot afford. At home Angélique practices the piano and also makes pretty rosettes from tricolor satin ribbons to pin to our dresses. Maman insists that we wear the tricolor cockades to show support for the revolution.

Sunday — July 8, 1792

What lies beyond the firmament? What alien worlds gravitate around other suns so far that I cannot see? A star near the edge of the night sky beckons me, urges me to seek. What should I search for? What is my destiny?

Perhaps it is the warm night or the quietness around me, but at this moment I do not wish to read or solve equations. Tonight I feel like looking up, gazing deeply far away, trying to decipher my destiny in the vast universe open before me.

Saturday — July 14, 1792

It's the second anniversary of the destruction of the Bastille. This time we did not celebrate. The violence and attempts against the king have turned Paris into a city of fear. Last week, the revolutionary government proclaimed that the *patrie est en danger*. Yes, we all are in danger, especially the royal family.

Paris resembles an armed camp. The black flag is flying over the Hôtel de Ville. Every day we see groups of soldiers parading in public places, heavily armed and chanting the *Ça Ira*, making me feel uneasy. It is like observing a dark sky full of clouds, foretelling a violent storm ready to erupt.

At home sometimes the tension makes us snap at each other. The other day Millie was cleaning Mother's bedroom, singing a catchy song, but Madame Morrell angrily asked her to stop. The chorus goes: "We will win, we will win, we will win. The people of this day never endingly sing. We will win, we will win, we will win. In spite of the traitors, all will succeed." Then I listened to another verse in the song: "Let's string up the aristocrats on the lamp posts!" It became clear. Millie was singing *Ça Ira*, a revolutionary song favorite among the sans-culottes to insult clergy, aristocrats, and the wealthy bourgeoisie. I am sure she hears the song in the streets so often that she sang it without even being aware what it means.

Monday — July 30, 1792

The soldiers from Marseille arrived in Paris this afternoon. Millie burst in out of breath to tell us. We raced to the open windows to catch the battalions of soldiers marching in, singing with loud tenor voices a patriotic hymn. People went out to give the soldiers flowers and food. Teenage boys scampered after the soldiers, cheering and laughing. Paris was in a festive mood, because the soldiers brought with their song a message of hope.

But the political situation is very contentious. Papa read a copy of the Brunswick Manifesto, a declaration written by the Duke of Brunswick, the commanding general of the Austro-Prussian Army. The duke warned the citizens of Paris to obey the king. The declaration is menacing; it threatens harsh punishment if people do not obey.

The Brunswick manifesto responds to those from the revolutionary groups who oppose Louis XVI, those who do not recognize him as sovereign of France. The Assembly was offended by the duke's declaration and ordered the sections of Paris to get ready for war. Yesterday, Maximilien Robespierre called for the removal of King Louis. Robespierre is a deputy to the Convention, a lawyer and leader of the most radical political group against the monarchy. After his speech, some Paris sections demanded dethronement.

Father is in his library with some colleagues, speculating what will happen next. The manifesto created both fear and anger among Parisians. The revolutionary leaders are inciting the people of France to revolt. I am afraid of the consequences.

Sunday — August 5, 1792

Violence intensifies. Millie came in this morning very agitated and hysterical, almost in tears, with disturbing news. A group of sans-culottes were ransacking the home of Comte Voyeaud and there was a big commotion in the streets. Millie described the looting and the pandemonium she witnessed. Mother had to give her a cup of tea to calm her down.

A large group composed of men and women ransacked the château, taking everything they could. Millie ran as fast as she could to tell us. The mob threw out the family's clothes, furniture, and paintings. She told us that many of the lawless men were drunk, shouting madly, and inviting people passing by to take the goods. We wondered about M. Voyeaud's family. And since they live just a few blocks away, I feared that the ruffians would expand their pillage to our street. Millie ran downstairs to secure the tricolor ribbons on all doors and windows.

Now the city is quiet. The bells of Notre-Dame just sounded the midnight hour. What plot is being conceived in the minds of the revolutionaries right now? Only time will tell.

Friday — August 10, 1792

The sounds of the alarm bells shattered the stillness of the summer night. I was in bed when the peal of the church bells roused me. I dashed to the open window, peering into the night while the bells sounded the call to arms. The tocsin was ringing and ringing, and soon the angry people of Paris would answer, marching against King Louis.

I rushed to join my parents when I heard voices and the footsteps of people gathering in the street below. Torches illuminated the stagger of those who answered the call, walking through the shadowed streets. The points of light seemed like a swarm of fireflies buzzing with rage, moving towards the Tuileries Palace where the king slept unaware of the danger. I shivered just hearing the sounds of enraged voices trailing away. I had a sense of foreboding in my chest, but I could not imagine what was to happen later.

My mother kept muttering, "*Mon Dieu! Mon Dieu!*" and as firmly as she could, sent us back to bed. I could not sleep, not even after the church bells stopped ringing and the voices in the distance became silent. Beneath the apparent quietness, fear enveloped the city. I was up long before the first light of dawn glimmered on the horizon, and I saw from my window a dark red sky, a view of the morning I had never seen before. It was a premonition of the blood that would flow hours later, tinting the streets deep red.

The storming of the Tuileries Palace, August 10, 1792.

After breakfast, while helping Maman fold the linen, the sudden firing of firearms in the distance startled us. The clatter was terrifying and I felt as if lightning hit me. My sister Angélique ran into the arms of our mother. The angry mob on rue Saint-Antoine was moving on their way to attack the Tuileries Palace.

It was terribly hot, but Mother kept the windows closed and the curtains drawn all day; the sounds from shooting and screams were not muffled by the still air. My father came early, saying the revolutionaries had taken the royal château. King Louis and his family fled to the Legislative Assembly building before the furious mob arrived at the Tuileries. A violent confrontation broke out across the gardens, with groups of angry citizens fighting the king's Swiss guards and the National Guards.

In the attack, the Swiss guards, the maids, and cooks of the palace were butchered. Papa described how the Champs-Élysées and the Tuileries gardens were covered with corpses and blood. We were trembling just imagining the horror. Maman was sickened, her face pale, repeating with shaky voice, "*Mon Dieu! Mon Dieu!*" The attack on the palace was brutal, as if all the hate and revenge boiled over, destroying any sense of decency in people. The king and his family were unharmed in the assault. However, they are now under house arrest.

Attack on the Palais-Royal — Insurrection of August 1792.

After dinner I went to Father's library. But I could not concentrate, feeling the tension of the city that lingered long after the attack. The massacre continued into the night, a horrific event that tormented my dreams. The fire that started earlier in the Tuileries Palace is still burning and the smoke seeps through the closed windows. How long can we bear this madness?

Tuesday — August 14, 1792

France is crushing its king. Today the leaders of the revolution publicly condemned King Louis-Auguste XVI and Queen Marie-Antoinette. The royal family is imprisoned in the tower of the Temple, including the children and the king's sister. The sans-culottes got their wish: to humiliate, defeat, and send Louis XVI to prison.

What has he done that is so horrible that the deposed king is thrown into a dark dungeon, where only the worst criminals belong? It's an injustice. I feel sorry for the king and his family; even if he was not an efficient ruler, this is no way to seek justice. Long ago, Queen Marie-Antoinette alienated the affections of the French citizens. Rumors circulated for years that she used to host extravagant fêtes while the peasants were starving just outside the gates of Versailles. Her enemies claimed that Marie-Antoinette thought only about gowns and jewels, spending enormous sums of money in frivolities and lavishing expensive gifts upon her lady friends.

Even if true, those excesses do not justify the punishment. I am sorry for the family, how can anybody not pity them? Princess Marie-Thérèse is only fourteen years old, just like my baby sister, and the charming Dauphin is only six. Children should not be thrown into a prison. What is their crime? Doesn't anybody in the government feel any compassion for them?

I wish there was a way to stop the violence that sweeps my beloved Paris. Arrests of innocent people have become commonplace. A few days ago, the Abbé Sicard, a teacher and protector of deaf-mute children, was imprisoned in the Abbaye along with many other priests. Students and teachers from the school have gone to the Assembly to plead for his freedom. What has he done to be arrested? Nothing. Papa says that his only infraction is being a priest who resists being ruled by the new government. Is this the revolutionaries' ideal of justice?

Thursday — August 23, 1792

The alarm bells awoke us. "The Prussian army is advancing!" yelled the town crier early this morning. The call to arms sounded and the city gates closed in haste. Drums beat all day, people ran in the streets like mad, and

The royal family taken prisoners to the Temple, August 13, 1792.

children wailed amidst the chaos. I saw through my bedroom window soldiers on horseback, trying to keep the peace in a frightened city. The smell of gunpowder sifted through the closed windows along with the hysteria of war. It was a scorching hot day, but the tension of the impending attack on the city was more oppressing than the air. We spent the long hours in a state of disquiet. I tried to distract my mind, finding refuge in my books, to see me through the rigors of the siege.

It is already past midnight and the city is quiet. But what terrible things are happening now, hidden under the darkness?

Saturday — August 25, 1792

Looking at the stars in the dark night, my mind wanders, taking me back to numbers. When I began to study prime numbers, I wondered if the number 1 belong in the set of prime numbers. There is a very important theorem which states: "Every whole number greater than 1 can be expressed as a product of prime numbers *in one and only one way*."

I take any number, and write it as the product of its factors. For example, the number 6 can be written as $6 = 2 \cdot 3$ and as $6 = 1 \cdot 2 \cdot 3$. According to the theorem, only the first expression is valid. This is the only one way to express 6 as the product of prime numbers. That explains why 1 is not considered a prime number.

Is there a way to prove the validity of this theorem for any number without having to check arithmetically one by one? For example, starting with a positive integer n, and I want to know if it is prime. To prove it, first I'd see if 1 divides n, and then check the reminder. I'd probably check dividing by any other integer until getting to n. Of course then I'd have $n/n = 1$. Well, this may be one way to formulate a proof, but perhaps it is better if I first study what mathematicians have done in this respect, just to be sure.

Thursday — August 30, 1792

Paris continues under siege. The Commune, the new revolutionary government, is using terror tactics to control the citizens. That includes authorized home searches. If the Commune suspects someone is a conspirator or a royalist, they send a group of soldiers to the home to arrest him or her. They call the searches "domiciliary visits," as if they were friendly calls, but in reality the sans-culottes hope to uncover traitors and conspirators who could threaten the nation.

Anybody could be targeted for a home search, even innocent people like us. The leaders of the Commune justify this intrusion as a necessary measure to seek out hidden firearms and apprehend suspects; they look for incriminating evidence to use against a citizen. The soldiers can show up late at night or in the early morning to ensure everybody is at home. The search party includes ten or more heavily armed men with sabers, pikes, and guns, and they rummage around a house, not leaving anything unturned. Most arrests so far are in the district Montagne Sainte-Geneviève, but I wonder how long before we see the search parties here. How do we know whose home will be next?

Sometimes I hear Papa and his friends carrying on their political debates, and they warn each other to keep quiet in public. One citizen can denounce another, even with little or no evidence. How many friendships have crumbled because of treachery? Everybody is afraid of everybody. I hope Papa has loyal friends who share his beliefs, and who will not betray him.

Father came this evening carrying a new book that M. Baillargeon was saving for me in his bookshop. It is written in Latin and it will be a struggle to understand it. So far I translated the title, *Philosophiae naturalis principia mathematica*, as "The Mathematical Principles of Natural Philosophy." M. Baillargeon told me that the gentlemen who purchased this book before called it simply the *Principia*. The author is Sir Isaac Newton, considered one of the greatest mathematicians of the last century. I find the subject rather difficult and the text hard to understand. The *Principia* contains so many concepts and ideas so new to me.

It's hard to remain calm in these uncertain times. Right now I feel a bit apprehensive about staying up late, afraid that the Commune might send a search party to my home tonight. What would I do if heavily armed soldiers would suddenly appear in my bedroom? Papa would not allow them to harm me, but what if he is arrested? Oh no, I do not even want to think about it. Perhaps it is best that I do my studies during the day.

Monday — September 3, 1792

A call to arms led to more massacres! About one o'clock yesterday afternoon the tocsin startled us. It was followed by gunfire and the beating of drums. A member of the Commune on horseback raced through the streets, proclaiming the enemy was at the city gates. The Prussian army had captured Verdun. We saw through the windows people running, assembling in different parts of the city, armed with pikes and pistols, with a gruesome anger showing on their faces.

Papa later described to Maman the horror that ensued after the alarm sounded. At about seven in the evening, enraged mobs surrounded the *Église des Carmen* and massacred all the priests incarcerated there. Then the wild crowd assaulted the prison of the Abbaye and killed more people.

My mother cried, hearing of such barbarity. She kept muttering, *Mon Dieu! Dieu le Père!* Even Papa, usually so calm and controlled, had to take a moment before concluding that he had no words to describe the shocking cruelties, the bloody spectacle, the madness of the people he was unfortunate to witness. "But why kill the priests?" Mother inquired. We knew that there was no acceptable answer to justify the slaughter. The sans-culottes believed that the prisoners would escape to help release the imprisoned King Louis and start a counterrevolution. Their anger and distrust of the monarchy has clouded their reason.

This morning, as Madame Morrell was assisting Mother with her coiffure, Millie came in, breathless and shaking with fear, to tell us that the

The massacres of the prisoners, September 2–5, 1792.

dead bodies from the church massacres were laid on the Pont-Neuf to be claimed. She had to walk by the bridge on her way to the market and witnessed the macabre spectacle. I closed my bedroom window, afraid to get a glimpse of what could only be a most gruesome nightmare unfolding so close to home. Mother immediately cancelled her visit to Madeleine and instructed Monsieur Morrell to drive Papa using another route away from the bridge. But it was too late; the coachman replied that they had already seen the horror Millie was describing.

The pealing of bells continues, echoing through the air, day and night, like desperate cries for help to the heavens. Is anyone there listening?

Saturday — September 8, 1792

Madame de Maillard came to visit this evening. When she arrived I was reading, struggling to translate a page in Newton's *Principia*. After she kissed my cheeks, Madame addressed me affectionately as "Sophie, the mathematician." It is the best compliment!

Glancing at the book before me, Madame sat to tell me the same book was translated into French, by a remarkable woman! She told me that over forty years ago Gabrielle Émilie le Tonnelier de Breteuil, Marquise du Châtelet, translated Newton's book. This remarkable lady scholar also wrote new sections that explain additions and corrections that were made later to the *Principia*. With her notes, the Marquise du Châtelet clarified many difficult concepts in the original book. Before Mme de Maillard finished our little interlude, I was already planning to search for the translation of Newton's book. Madame added a bit more about the Marquise du Châtelet. Here I summarize what she related to me.

Gabrielle Émilie le Tonnelier de Breteuil was born in 1706 in Paris. Her father, the Baron de Breteuil, was principal secretary and introducer of ambassadors to Louis XIV. As a child, Gabrielle Émilie had tutors, including her father who instructed her in Latin. Mlle de Breteuil was very intelligent and had a high aptitude for languages, mathematics, and the sciences. By the time she was twelve years old she could read, write, and speak fluent German, Latin, and Greek. At age nineteen she married the Marquis du Châtelet and had three children.

Later, Madame du Châtelet hired tutors to teach her geometry, algebra, calculus, and sciences. She was interested in the work of Isaac Newton, so she asked mathematician Moreau de Maupertuis to teach her the theories of

Newton. An eminent mathematician himself, Maupertuis supported Newton's ideas that were, at the time, debated by French scholars.

Émilie du Châtelet is also remembered for her tenacity. She wanted to attend the regular meetings of the Academy of Sciences, where scholars discussed the latest scientific discoveries. However, women were not allowed to attend these gatherings. Thus, she had to meet with colleagues at a coffeehouse.

But women were also banned from those establishments, including the popular Gradot's Café, where philosophers and scholars met regularly to exchange ideas. To exclude women, the owners ruled that no person wearing skirts was allowed in their coffeehouses. Thus, Mme du Châtelet dressed in men's garments entered the coffeehouse and sat with Maupertuis and other scholars. Allegedly, the gentlemen ordered a cup of coffee for her, treating the lady as another member of the group. The proprietors pretended not to notice that they were serving a woman, and from then on Mme du Châtelet could walk in the café and nobody bothered her.

When she was twenty-eight years old, Mme du Châtelet met the philosopher Voltaire. The two moved to her Château de Cirey to write about science and philosophy. Their collaboration was as close as their friendship. In the *Introduction to the Elements of the Philosophy of Newton*, Voltaire stated that he and Mme du Châtelet worked together in the writing of the book. After this joint project, Émilie du Châtelet continued her study of mathematics and completed her translation of Newton's *Principia*. Alas, she died before her translation was published.

It is a fascinating story. I am so glad I had a chance to learn about an incredible woman scholar. But now it is important that I find Châtelet's translated version of Newton's *Principia*. Monsieur Baillargeon must not have a copy, or he would have told me already. I will ask Papa to help me locate it in another bookstore.

Monday — September 10, 1792

Violent conflicts erupt in the streets of Paris every day. We live in terror. The gates to the city are closed, armed soldiers patrol the streets, houses are searched for weapons, and the fear of persecution is felt everywhere. Maman has placed tricolor ribbons across doors and windows "to keep away the bloodthirsty crowds roaming the streets," as she says. The tricolor ribbons are a symbol of our loyalty to the republic; we are hoping such display will keep us safe from the fanatic revolutionaries. We spend most of the time on the upper floors, fearing for my father when he goes out to tend his business. Millie's errands are limited to the essential trip to the bakery and the market.

More attacks on the prisons continued after the first massacre of prisoners. It is not just the priests in the prisons who have fallen victim to the angry mobs. Many other citizens are murdered as well. At the Bicêtre prison many of the people killed were younger than eighteen years of age! Father says that most of the killings are preceded by a trial, run by a mob court devoid of any justice. The so-called judges in these courts are the actual killers themselves, and most of the time they are drunk. The sight of the killers described is revolting — covered in blood and laughing.

The most hideous killing was that of Princesse de Lamballe, the dearest friend of Marie-Antoinette and former superintendent of the queen's household. She was imprisoned just for being close to the queen. During her trial the young lady refused to denounce King Louis and Queen Marie-Antoinette. This enraged her accusers. Then the princess was stripped, raped, and her body was mutilated. Why? What could possibly cause such a horrible monstrosity? I don't want to imagine the hideous tortures inflicted upon the young woman by the savage multitude.

My father blames the massacre on the fear of a counterrevolution. The revolutionaries defend Paris by watching for rebels and traitors that oppose the revolution. But why do they massacre people? Why such inhumane brutality? The Princesse de Lamballe was innocent, and yet they killed her like an animal, and for what? What is the murderers' justification?

Saturday — September 15, 1792

Émilie du Châtelet must have been a great woman. I envy her strength and bold spirit. She was privileged to have had tutors and opportunities to pursue her intellectual ambitions. But she also had to cope with the prejudice of a society that did not take women scholars seriously. I am impressed by the range of her accomplishments. She translated into French several classical works written in Latin and other foreign languages. For example, she translated *Oedipus Rex*, a play by Sophocles written in Greek.

Madame du Châtelet also translated *Institutions de physique*, which was an explanation of the metaphysical theories of the German mathematician Gottfried Wilheim von Liebniz as expressed in his *Monadologie*. And of course Émilie du Châtelet's greatest contribution to France is her translation of the *Principia* by Isaac Newton from the original Latin into French. I can't wait to read it.

Thursday — September 20, 1792

Finally Papa found the French translated version of Newton's book the *Principia*. I read a full chapter, but I still cannot understand much of it. So far I can only summarize some of Newton's ideas. He begins by defining the concepts of mass, motion, and three types of forces: inertial, impressed, and centripetal. Newton also gives definitions of absolute time, space, and motion, offering evidence for the existence of absolute space and motion.

Newton proposes "three laws of motion," with consequences derived from them. The remainder of the *Principia* continues in the form of propositions, lemmas, corollaries, and scholia that are quite difficult to comprehend. Book One, *Of the Motion of Bodies*, applies the laws of motion to the behaviour of bodies in various orbits. Book Two continues with the motion of resisted bodies in fluids, and with the behaviour of fluids themselves. In Book Three, *The System of the World*, Newton applies the "law of universal gravitation" to the motion of planets, moons, and comets within the solar system. He explains a diversity of phenomena from this unifying concept, including the behaviour of the tides, the precession of the equinoxes, and the irregularities in the moon's orbit.

Even reading the translated version of the *Principia* I find it difficult to grasp some concepts, but I've made a commitment to study it. I intend to learn Newton's science one day.

Sunday — September 30, 1792

I am depressed. Although I try not to think about the massacres of the past several weeks, I also worry about Papa. He and his colleagues talk about the increase of violence and the arrest of innocent people charged as counterrevolutionaries. They say the prisons are full of falsely accused citizens. I try not to listen, but it is difficult to remain oblivious to the social and political horrors happening every day.

To brighten my mood Papa brought me a new book with the title *Lessons on Differential Calculus and Integral Calculus*. It was written by Monsieur J.A.J. Cousin, an acquaintance of my father. I hope the concepts of calculus in this book will help me understand Newton. The *Principia* is rather difficult!

Calculus deals with the rate of change of quantities, which I interpret as slopes of curves, and the length, area, and volume of objects. This book divides calculus into differential and integral calculus. The first deals with derivatives of functions, and the second with integrals of functions. I have to study calculus as earnestly and diligently as I have studied other topics. Before I begin, it would be interesting to learn what led mathematicians to develop this branch of mathematics.

Saturday — October 6, 1792

I have a new book on calculus written by Maria Gaetana Agnesi with the title *Instituzioni analitiche ad uso della gioventù italiana*. Papa explained that he bought this book because the Marquis de Condorcet recommended it for me. It has many examples to illustrate the methods of calculus. M. Condorcet and other scholars have praised it as the best instruction book in differential calculus.

Maria Gaetana Agnesi is an Italian mathematician. In the preface of her book, Agnesi thanks a person named Rampinelli for helping her learn mathematics. She wrote: "With all the study, sustained by the strongest inclination towards mathematics, that I forced myself to devote to it on my own, I should have become altogether tangled in the great labyrinth of insuperable difficulty, had not [Rampinelli's] secure guidance and wise direction led me forth from it . . . ; to him I owe deeply all advances (whatever they might be) that my small talent has sufficed to make."

Madame Agnesi was so fortunate to have had M. Rampinelli to teach and guide her studies.

Saturday — October 13, 1792

My parents invited a number of friends for dinner. They talked about the latest political events and argued about the new decrees imposed by the government. When I spoke to Madame Geoffroy, several voices in unison urged me to address her as *citoyenne Geoffroy*. At first I was perplexed and was not sure I understood correctly, but Papa then explained that from now on the titles monsieur and madame will be replaced by the appellations *citoyen* and *citoyenne*. According to him, it was officially decided by a resolution issued by the Paris Commune last week. Monsieur LeBlanc resumed the discussion, adding that the titles madame, mademoiselle, and monsieur are no longer acceptable. The new forms citizen and citizeness are intended to erase the distinctions between the various classes of society.

It makes no sense. The new form of address is insufficient when it comes to women. If a person says, for example, "Citoyenne Germain," who would know whether they are talking to my mother, a married woman, or addressing me? I do not dispute the use of "citoyen" to address males, after all, the social title "monsieur" is applied to all men, but I disagree with the use of "citoyenne," for it does not apply to all women. So, as far as I am concerned, my mother always will be "Madame Germain," and I always will be "Mademoiselle Germain."

I must ask, furthermore, how can anybody presume that abolishing courtesy titles makes us all equal? The new appellation rules will not change how people treat one another. People will always have prejudices, whether we address each other with title or without.

Monday — October 15, 1792

Paris continues under siege. Violence, terror, and scarcity of goods punctuate our lives. Tallow is in short supply. We are allowed only a few candles per week. Now I have to study by daylight. Sometimes I do not realize how dark it is, until the shadows of the night make the letters in my books disappear and my eyes strain to see. Millie and Angélique offered to give me their own candles, but it would be selfish on my part to deprive them of light, especially now that the nights are getting longer.

Saturday — October 20, 1792

Now I learn calculus. This is an area of mathematics much more challenging. I read somewhere that calculus was developed by mathematicians who had the need to explain physical theories. My research suggests that two scholars in different parts of the world came up with similar ideas simultaneously. But the way each scholar arrived at their calculus is so different.

It is assumed that Isaac Newton developed his ideas of calculus when the plague closed the schools in the summer of 1665. Newton returned to his home in Lincolnshire and, in seclusion, he began to lay the foundations of the new mathematics. The "method of fluxions," as he termed his calculus, was connected with the study of infinite series. A "fluxion," expressed by a dot placed over a letter, such as $\dot{x}$, was a finite value; for example a velocity denoted by $\dot{x} = \frac{dx}{dt}$, where x represents distance and t represents time. The letters without the dot represented "fluents."

The method of fluxions was developed on Newton's insight that the integration of a function is merely the inverse procedure to differentiating it. Taking differentiation as the basic operation, Newton produced simple analytical methods that unified many separate techniques previously developed to solve apparently unrelated problems, such as finding areas, tangents, the lengths of curves, and the maxima and minima of functions. Because his new mathematical ideas were so different, Newton feared criticism so he did not publish his memoir until 1704. It appeared as an appendix to his book on *Opticks*. Newton claimed to have written *De methodis serierum et fluxionum* in 1671, but he did not publish it until much later.

The German mathematician Gottfried Wilhelm von Leibniz had similar ideas. Leibniz corresponded with French mathematicians, and one day he came to Paris to work at the Academy of Sciences. It was during this period in France that Leibniz developed the basic features of his version of calculus. In 1673, he was still trying to develop a good notation for his calculus. A few years later he wrote a manuscript using the integral $\int f(x)\,dx$ notation for the first time. In the same manuscript he gave the product rule for differentiation.

By 1676, Leibniz discovered that $d(x^n) = nx^{n-1}dx$ for both integral and fractional n. For some unknown reason Leibniz did not publish his new mathematics. In 1684, he published his differential calculus in a memoir entitled *Nova methodus pro maximis et minimis, itemque tangentibus* (*A New Method for Maxima and Minima as well as Tangents*), published in *Acta eruditorum*, a journal he founded in Leipzig two years earlier. The paper contained the dx, dy notation, the rules of differentiation, including $d(uv) = udv + vdu$, and rules for computing the derivatives of powers, products, and quotients.

Mathematicians in France now use Leibniz's notation — derivatives expressed as dy/dx — rather than the fluxions developed by Newton. Leibniz was also the one who called this branch of mathematics *calculus differentialis* and *calculus integralis*. In 1696, the first textbook on calculus appeared. The *Analyse des infiniment petits* was written by the French mathematician Marquis de l'Hôpital, who had learned calculus from the great Swiss scholar and mathematician Jean Bernoulli.

Perhaps one reason why I have had difficulty understanding Newton's book is due to notation. I will ask Papa to help me find any publications by M. Leibniz. Of course the textbook by l'Hôpital may be even better since it is written in my own language. I must send a message to M. Baillergon to locate this book for me.

Friday — November 2, 1792

I wonder about the future of my beloved France. The brutal killings and the social chaos have changed my country. Right now there are two competing views on which direction France should go, embodied by two political parties: the moderate Girondists who favor a peaceful reconstruction of France, and the more radical Jacobins, led by Maximilien Robespierre who favors purging France of its imperial past. Who is right?

Meanwhile at home we try to maintain a sense of normality. We go about our lives as if nothing has happened, but I know this is only superficial. We all have witnessed the horrors brought about by the misery of the poor, and the misguided eloquence of the political leaders. Even my little sister has seen death in the face; she no longer runs to Mother when the mobs are heard outside our windows, searching for a victim, searching for justice and a truth that seems to elude them.

We all have grown beyond our years. Mother's beautiful face is lined, and a mask of worry and apprehension seems to hang over her eyes. Her smile is not as easy as it was years ago. My parents no longer go to the theater, and we visit only our closest friends. But we try to maintain our family traditions. We have our literary evenings and read plays by Molière and passages from Rabelais. Every week Angélique regales us with a beautiful piano concerto, and Mother sings her beloved arias. Millie joins us at these gatherings. She now reads so well and is always eager to recite popular poems.

I have my books to see me through the rigours of a siege that doesn't end. My books, they are quietly waiting to reveal the knowledge imprinted in their pages, opening a door to a world without violence.

Thursday — November 15, 1792

King Louis is ill. The newspaper reported that his health is quickly deteriorating after living imprisoned in that dreadful cold prison. Our morning prayers were dedicated to him. My mother recited the seven psalms for his recovery. I am still confused, trying to understand why the king and his family are prisoners in the Temple. What is their crime? What about the children?

I am glad that my father is no longer a deputy in the government. Papa resigned long ago, even before the royal family was locked up, because he did not want to share the blame for crimes and follies, which he does not condone. After so much injustice and violence, Papa had no option but to leave the government.

Saturday — November 17, 1792

I have struggled to understand the concept of the differential dx. If Δx represents a distance along the x-axis, or the difference between any two values of x, dx means exactly the same thing, with one difference: dx is a differential distance, a very, very tiny difference. If I have a quantity whose value is virtually zero, then $2 + dx$ is, well, 2. Or if I divide a number by dx, such as $3/dx$, the result is infinity!

There are two cases under which terms involving dx can yield a finite number. One case occurs when I divide two differentials. If they are the same, we get the quotient equal to 1, such as $dx/dx = 1$. However, if the differentials are different, the quotient can be some finite number since the top and the bottom are both close to zero, such as $dy/dx = $ constant, or any value. The other case is when I add up an infinite number of differentials, the sum is equal to some number greater than zero and less than infinity. It makes sense. These two cases describe the derivative and the integral, respectively.

Eh bien. The flame of my candle flickers nervously, casting shadows on the walls like ghosts and apparitions. I must bid good night before the flame expires and leaves the room in total darkness.

Sunday — November 25, 1792

The health of the king is worsening. And yet, he remains imprisoned in the Temple, deprived of all bodily comforts. What is going to happen to him and his family? Father and his friends consider whether Louis-Auguste XVI can be tried in court. Many of the revolutionary leaders have accused him of crimes against the nation. However, the Constitution of 1791 protects the monarch from any penalty worse than dethronement, and no court in the land has legitimate jurisdiction over the king.

Father is not sure there are enough citizens in Paris to use this argument anymore. All his friends insist that King Louis cannot be condemned without a trial. Of course, Papa is on the side of the monarch, but he feels that, with no other alternative before them, the Convention may just take on the role of a court. I do not understand; this is contrary to accepted judicial principles. I do not comprehend the law, but surely this does not look good. Mother keeps praying, hopeful that somehow this nightmare will end. I am afraid the revolution has gone past the point of no return.

Saturday — December 1, 1792

I am furious! Mother called me to her sitting room to give me a lecture on manners. What made me angry is that she told me to be especially courteous with Antoine-August, who will come for dinner with his parents next week. Maman reminded me to behave like a nice young lady. For Maman, being a lady translates into conducting oneself like a naïve, helpless woman who allows a man to feel superior. I can't do that! Why are women expected to be the weaker sex and be subservient to men?

I don't wish to talk with Antoine-August anymore. He has become conceited, condescending, patronizing, and acts as if he knows everything. With his ruffled shirt-fronts and lace at his wrists he seems to have spent hours dressing. Last time I saw him he wore a white embroidered waistcoat turned back with blue. His hair was so powdered that traces of white lingered in the air around him.

But what annoys me the most is that he intersperses mathematical terms in his speech to appear as if he knows a lot. But he doesn't fool me. Once I tried to correct him when he stated something so absurd that even I who am not a trained mathematician knew better, but he acted as if I was the stupid one, and he stopped me before I could make him aware of his error. I was so angry that I didn't have the words to debate him. Antoine-August can impress people with his airs of intellectual prowess and his sophisticated words; he is so arrogant! Sometimes he speaks nonsense; once, he even confused irrational numbers with fractions! But he will never admit it.

After Mother's lecture I was even more rattled because she also selected for me the brown dress to wear at the party. I dislike that dress! I was sulking in my room when Millie came in, and to console me she sang a song that says: "*Il faut bien que je supporte deux ou trois chenilles si je veux connaître les papillons.*" After a few minutes, I had to smile at Millie's silly song. Perhaps she is trying to tell me something. Maybe it would benefit me to be friends with Antoine-August. He could tell me about the topics he is studying at school, especially if he thinks I would be impressed. Possibly he would let me see his lecture notes in mathematics. I will be polite as long as he does not treat me condescendingly. I set a limit on that.

Friday — December 7, 1792

Why don't women demand an education like men do? Boys have tutors to teach them science and mathematics, and then they go to the schools of higher learning to pursue degrees. As scholars men correspond with others in their fields of research, and learn from one another. Men can become lawyers, doctors, mathematicians, or anything they want to be. However, women are excluded from these professions; they stay home and raise children. Even if we are smart and wish to learn sciences and mathematics, we are told it is a right reserved to men only; we are expected to be content knowing only the fine arts and literature, but some areas of knowledge are "not suitable for women." Why?

When I solve an equation or prove a theorem, I feel a sense of accomplishment, an intellectual pleasure, but I am still Sophie, a girl. If I can think mathematics, then it means mathematics is not reserved for the minds of men only. Who decided to exclude women from this pursuit? Women have demonstrated that they are as intelligent as men. Many women in history have made important contributions to the scientific development, starting with Hypatia of Alexandria in ancient Greece. I know of two other women scholars in this century, Émilie du Châtelet and Maria Gaetana Agnesi. There must be many more women as intelligent as these women, and I am sure now there are many more like me who aspire to scientific learning, but we are repressed by society and excluded from schools.

Oh, I would give anything to have the privilege to study with a mathematician. I have learned so much in the past three years, but not enough! Because, even though I have struggled to feed the hunger of my intellect, I have become fully aware of my status as a woman in our society. The servitude and prejudice under which society keeps women oppressed is humiliating. I hope that the new Republic of France will justify itself by offering women opportunities for education equal to men's. The social demands of the people were heard but, as of today, education is still not a right of women.

Friday — December 14, 1792

I am anxiously waiting for a letter from Antoine-August LeBlanc. He and his parents came to dinner a week ago, and he promised to send me some of his class notes. I must say, I did not expect this turn of events. At the party, as always, the conversation began with politics and the state of affairs in Paris. It continued with discussions about the royal family imprisoned in the Temple, and the impending trial of the king. Antoine, at seventeen, is now an adult, so he spent the evening arguing with the gentlemen about the case, expressing his fervor for the republic.

I lingered for the opportunity to talk with him because I had so many questions to ask about his studies. It wasn't until midnight, when they were about to leave, that I gathered the courage to approach him. He was surprisingly attentive to my questions, quite courteous, and he even offered to send me copies of his lecture notes. He mentioned that he is studying differential equations. He claimed that by knowing calculus one could understand the world. Well, I wish to know it too. And I shall learn calculus as well.

The following day I went to the bookshop and asked Monsieur Baillargeon for books on differential equations. He found several used books written by Leonhard Euler. I was delighted and without a second thought I bought them all. One of the books is titled *Institutiones calculi differentialis*. The other is a three-volume book, *Institutionum calculi integralis*. Now I have the best books to learn differential and integral calculus.

Euler included a theory of differential equations, Taylor's theorem with many applications, and many more concepts I am eager to learn. I have just read the section on differential equations where Euler distinguished between "linear," "exact," and "homogeneous" differential equations. I spent the day reviewing functions, which Euler presented in a clear manner. A linear differential equation is of the form $dy/dx + P(x)y = Q(x)$, where $P(x)$ and $Q(x)$ are continuous functions (of x). This type of equation is easy to solve by integration. Exact differential equations look like this: $M(x, y)dx + N(x, y)dy = 0$, where M and N are functions of two variables (x, y). The method of solution is still not clear to me. Thus I need to devote more time to practice solving this type of equation.

Now I am beginning to understand this branch of mathematics. Calculus can be useful for understanding how things change, or more precisely to study "instantaneous" change, over tiny intervals of time. The theory of differential equations has an inherent beauty because it provides us with a unique tool for understanding nature. There is so much more I need to learn, and that is why I am determined to master calculus by next spring.

Tonight I look up at the sky; there is no moon in sight. It's past midnight already. The only sound I hear is the howl of the wind outside my bedroom window. I think about the king, the queen, and their children, locked up in that cold dark prison. What future awaits them?

Thursday — December 20, 1792

Paris has changed so much. When I was a little girl I looked forward to this time of year full of joyful anticipation. The white snow, the warming fires, the glow of lanterns and the preparation for Christmas; the prelude celebrations transformed Paris into a city of light and merriment. We used to go to church, and kneel under the radiance of a thousand candles. I enjoyed so much the pastoral festivals that announced the approaching celebration of the Nativity of Jesus. Even the poorest citizens would look at the holidays with hope, and the wealthy patrons were more generous, giving away presents to those who had none. Those blissful days are gone.

The cold weather is here again, but this time no fire can warm the hearts of the people. Something changed since the day the monarch of France was arrested as a common criminal. Now his trial is unfolding as a mockery to justice. My mother prays for a miracle, but the king's lawyers cannot defend him against the terrible accusations.

The trial is a parody. Many people have already condemned him. Why doesn't anyone shout that this trial is illegal? King Louis is charged with treason. If the jury finds him guilty, the legal penalty is death. Some of Father's colleagues argue that to allow the king to live would undermine the principle of revolutionary justice. What justice? Murder is committed in the name of justice. No, there is no justification for the bloodshed that now soils the ground of my beloved Paris.

I fear for my father. Although he stands for change and supports the revolution, he sympathizes with the king. If he would vote, he would not condemn Louis XVI to die. I am sure of that. My father is outspoken, not afraid to stand for what is right. But there are many hypocrites who betray their friends, even their own brothers. My mother fears that one day somebody may accuse my father. But, as he says, there is no sense in worrying about what has not happened. And he goes on fearlessly speaking against injustice.

Tuesday — December 25, 1792

I had a wonderful dream last night. When I woke up I felt so happy, with the remains of the vision still lingering in my mind. It was a vivid dream, like a revelation. I was floating, flying far away from the Earth, where the darkness of the sky was illuminated only by the millions of stars shimmering in the distance. My body felt like a feather, weightless and free. As I soared, I spotted mathematical equations and symbols floating in all directions towards and away from me, appearing and disintegrating like soap bubbles.

There were letters x, y, z, and numbers $\pi, e, i, 1$, dancing like butterflies in a flower garden. When I'd reach out to take one equation, it would fly away farther from me and then another would appear near by. There were many kinds of equations, some simple ones I recognized, but other equations were complicated and I did not understand them. There was one particular equation that I needed to have, so I chased it with all my might. When I held the equation in my hand, I felt much joy, a happiness that went beyond laughter. The equation was not a real object, but it felt like a deep indescribable emotion; although I did not touch it, I sensed it, making my heart beat fast.

Upon awakening I could still feel the euphoria. My hand was still holding the book I was reading last night. I got up quickly, trying to remember the equation because I wanted to write it down to try to understand its significance, but I could not recall it. The vision of the equation was gone. What remained in my mind was just the feeling of sheer pleasure I felt holding the exquisite equation.

I am thinking now, if I could conceive an equation while dreaming, it must reside somewhere in my head. Thus, one day the cherished equation will reappear, not in a dream but in my awakened mind.

Paris, France
1793

⁕

5

Upon the Threshold

Monday — January 7, 1793

A new year begins, a sickly child born to a mother cruelly beaten, crushed. The mother is France in 1792, a year that began in strife and ended in sorrow. A year ago there was suffering, people rioting for food, dying of hunger and despair. Then the country was besieged by war.

The summer was no better; the citizens of Paris went crazy and stormed the Tuileries Palace, taking the king and the royal family captive and throwing them in a cold, dark prison. The people spat on the faces of the monarchs and their children, treating them all like criminals. Violence, rioting, and more massacres followed. In September bloodthirsty mobs murdered hundreds of innocent people, including priests imprisoned solely because they did not swear allegiance to a civil constitution. Overnight the face of Paris changed. A radical Commune took over the government, and the gates to the city closed.

The trial of the King of France filled our hearts with trepidation. Louis-Auguste XVI was condemned like a common felon, judged in a courtroom full of disloyal subjects, people who turned against a monarch who was once considered the father of the nation. Today Louis XVI was declared guilty. His lawyers could not convince the jury that King Louis meant no harm to the nation. His accusers will ask for the death penalty.

A new year sneaked up on us without fanfare, as subtle as time when day becomes night; I hardly noticed it. This time there were no joyful celebrations, no gifts, no feasts, no laughing. We just wait, anxiously, hoping that the life of the king will be spared.

Wednesday — January 16, 1793

The wait is over. Today our beloved King of France Louis-Auguste XVI was condemned to die. The decent people of Paris are at a loss for words. The radical revolutionaries are the only ones who rejoice. We are stunned. For weeks Mother prayed for King Louis, hopeful that he'd be found innocent or given a lenient sentence. We kept anxiously waiting for news, awakening each day wondering when the nightmare of France would end.

Mother wept when Father told us. Even Angélique stayed in her room drawing, trying to immortalize the present with her charcoal sketches. We dread what will happen next.

Father and his friends reviewed the conclusion of the king's trial. The voting process took several days to complete; only a few delegates abstained from voting. At the first vote, 693 deputies unanimously voted the king guilty, and the call for a referendum was rejected by 424 to 283. The death penalty carried 387 to 334 votes. Seventy-two deputies asked for a reprieve, but an extra vote saw their demand rejected by 380 to 310. On each occasion deputies answered individually to their names. That is how we know that Monsieur Condorcet voted against the execution of Louis-Auguste XVI. Why weren't more people like him? Mother continues hoping, reciting her prayers for a miracle that could save King Louis. Papa says it is already too late for prayers.

My heart feels oppressed, my spirit trembles. I must find refuge in my studies. Mathematics is a world without violence where I can thrive and flourish, untouched by the grim spectre of murder and sorrow.

Monday — January 21, 1793

Today is one of the most reprehensible and shameful days for France. Today our King Louis-Auguste XVI was put to death at the guillotine. I must record this day for posterity.

We woke up dispirited knowing that something ghastly was soon to happen. A gray and cold sky greeted us, making me feel more miserable. No one at the breakfast table spoke a word and I didn't feel like eating. Even my sister Angélique was unusually withdrawn. The gates of Paris were closed earlier in the morning; the city was eerily quiet, in an expectant state, as if the crime that was about to be committed was insulated from the rest of the world by a wall of silence.

Later, Papa told us that before his execution, when the drums ceased for a moment, the king addressed the mob: "I die innocent; I pardon my enemies." But soon the beating of the drums drowned his voice and the heavy blade swiftly descended. Papa said with a low voice almost inaudible that the crowds cheered when the head of Louis XVI dropped to the muddy floor. Maman did not want to hear another word about it and left the dining room. I heard her weeping in her chamber, trying to intersperse a prayer between sobs. Maman did not bother to hide her sorrow when my sister and I approached her. The cold rain mimicked the tears that glistened in her eyes. I kissed her cheeks, moist with the salty drops. My sister clung to her, and I left her side to be alone.

Standing in the middle of my bedroom, alone in darkness, I began to cry like a child. I tried not to think, for I did not want to imagine the horror at the scaffold. My tears felt warm on my face; my throat hurt as I tried to muffle a sob. Why? I cannot comprehend the violence and the murder. Why did they have to take the life of a man in that horrendous manner? Why could they not spare him? What kind of justice is this when, after killing the king, people laugh and cheer?

Papa offered a hopeful attempt to comfort us: "A republic founded on the blood of innocent victims cannot last long." My God, I hope he is right and that this nightmare ends soon. My heart feels oppressed and my spirit heavy with sadness.

The life of the city has resumed its course after the execution of our monarch, and many shops opened as usual. But to me, it felt as if we had just buried a member of our family. My father stayed home. We just could not suppress our grief, and we mourned in private the murder of King Louis-Auguste XVI.

Friday — February 1, 1793

I have pneumonia. My chest hurts when I breathe and my body feels weak. Maman claims I contracted the illness for staying up late at night, studying in the freezing cold. It's possible my bedroom became cold when the fire in my heater was extinguished, and I did not pay attention.

Angélique stays with me most afternoons drawing pictures. She says that I had a fever so high I was delirious, and that I was mumbling strange words nobody could understand. She claims that once I was desperately reaching out for my books, and Papa had to hold me to calm me down. But I do not remember any of that.

The fever has subsided and I am beginning to feel better, just weak from being in bed for so many days. The doctor came this morning and gave me a bitter tonic difficult to swallow. Maman wants me to stay in bed a little longer and has forbidden me to read. Papa asked me to heed Maman's advice by not immersing myself so deeply in my studies that I jeopardize my health. He reminded me that if my body is weak by illness, I would not sustain the strength of my intellect. He kissed my cheeks ever so gently. I've promised him that I will not get up until I feel better.

Millie brought me chicken broth this evening and tried to cheer me up by reciting a love poem she learned, she said, while sitting by my bedside watching over me when I was very sick. She now reads so well, I am very proud of her. Then she told me that a letter addressed to me came in Monday's post from Antoine LeBlanc. My heart began to beat fast, thinking that the letter must contain the lecture notes on mathematics Antoine promised to send me. I asked Maman, but she replied that I must wait until the doctor says I am cured. But I cannot wait. I must get better soon. I've neglected my studies too long.

Wednesday — February 6, 1793

Madame de Maillard came to visit me. She brought me a book titled *Mécanique analytique* written by Joseph-Louis Lagrange, a leading mathematician in Paris. She says that it is a book for *real* mathematicians. I perused it, but I don't know enough mathematics to comprehend this material. All the pages of the book are full of equations. I must admit, I feel quite intimidated by what Lagrange calls "analytic mechanics." He divides the book into two parts: *la Statique ou la Théorie de l'Equilibre, et la Dynamique ou la Théorie du Mouvement* (*Statics or the theory of equilibrium, and Dynamics or the theory of movement*) and states that he will address separately solid bodies and fluids.

Monsieur Lagrange wrote in the Preface of his book: "One will not find figures in this work. The methods that I expound require neither constructions, nor geometrical or mechanical arguments, but only algebraic operations, subject to a regular and uniform course." And indeed, I perused the book and found just equations, not a single figure! But the equations tell so much if one understands their meaning. I'm sure one day I shall learn enough mathematics to understand this book. If Mme de Maillard thinks I can do it, then I will.

Monsieur Lagrange also states that for the study of *Mécanique analytique* he will use calculus and that he will introduce a novel demonstration of maxima and minima and mentions solutions to isoperimetric problems. Well, I will strive to learn calculus, so that I can better understand this unique book that must be very important for my education.

I've reviewed the notes that Antoine LeBlanc sent me recently. It is the same material that I found in Agnesi's book on differential calculus. Since I mentioned my interest in prime numbers, he also recommended that I obtain a copy of Adrien-Marie Legendre's memoir *Recherches d'analyse indéterminée*, published in 1785. He emphasized that this paper contains a number of important results such as the law of quadratic reciprocity for residues, and the results that every arithmetic series with the first term co-prime to the common difference contains an infinite number of primes.

This is quite advanced material for a lowly pupil like me, one that is deprived of tuition. Could I ever reach the summit of my aspiration to become a mathematician some day? If I can ever understand the material in these scholarly works, then I will categorically say I am a mathematician.

Monday — February 18, 1793

Paris has changed but my life has changed even more. I was a child when the first signs of the revolution manifested and now I feel so grown. My development is not the chronological effect of time that makes me see my existence differently. It's the social changes all around me. In only a few years, French society was transformed; the old customs and traditions are gone, and with them so many values that we held dear. Even the way we address each other has changed.

King Louis XVI is already forgotten; his memory is now but a shadow that follows the members of the royal family who remain imprisoned. They were put away, out of sight, as if the murderers are afraid to see the specters of their sins. Just like Marie-Antoinette, the once beautiful queen of France, the graces of human kindness have been discarded and replaced by rough, crude discourse. The monarchy was erased in one stroke, like multiplying by zero.

The decent citizens of France now live in fear. People are afraid of being accused, condemned, imprisoned, or sentenced to death. Many times I pondered what I would do without my studies. I could not bear the harsh reality without the researches in mathematics that take my mind into a world where violence and hate do not exist. Learning provides me with an escape; books soothe my mind and comfort me when anguish oppresses my heart not knowing if my father is in danger. How anxiously we live awaiting his return every evening.

It is terribly cold. The sun shone at midday but not long enough to melt the icy snow on the streets. We now spend more time in the kitchen near the burning stove because there is not enough fuel to replenish all the heaters in the house. Coal is scarce, and lamp oil is terribly expensive. Poverty is rampant in the streets of Paris. Maman and Angélique are knitting scarves for the homeless children, the orphans of the revolution.

Monsieur and Madame LeBlanc came to visit this evening. M. LeBlanc took the conversation away from politics and focused on education. He expressed the dire need to establish a new university to educate scientists and engineers. He indicated that the revolution has depleted France of intellectual power since many engineers and scientists have emigrated to other countries. Besides, he stated, the great schools in France are now closed, and the ravages of war have left the transport infrastructure in need of maintenance and improvement.

M. LeBlanc shared the discussions he had with a group of scholars to establish a unique school, where great mathematicians like Lagrange and Monge can train civilian and military engineers. He mentioned other names and details about the proposals for the new polytechnic school, including what kind of courses would be taught in mathematics, physics, and chemistry. Mme LeBlanc said that Antoine should be one of the first pupils at this school. He needs to be trained by the best teachers since he would like to attend the École des Ponts et Chaussées, an excellent engineering school that requires rigorous training in mathematics and science for admission.

Nobody mentioned admitting women to this school, but since the new Constitution of France is supposed to guarantee equality among all citizens, I wonder if someone like M. Condorcet again would raise the issue of women's education. I do hope the new government will make allowances for women to be admitted to the polytechnic school.

Nothing is for sure yet since the Academy of Sciences is closed and there is great confusion about the future of the new republic. I wonder if the plan is carried out, perhaps it will open a possibility for me. Wouldn't it be grand if I could enroll? I would do anything to be taught by the greatest mathematicians in France.

I must study hard because, as M. LeBlanc emphasized, the new school will admit only the most gifted students, and they will ensure that by requiring that students pass a rigorous entrance examination. I shall prepare myself. Oh yes I will. What else can I do to make my dream come true?

Wednesday — February 27, 1793

I feel strong enough to resume my studies. My mind is clear again to meet the challenges of a new topic that at first seemed insurmountable. I resumed my studies of differential calculus.

There is something magical about *Infiniment petits*. I went back to the basic definition: "a derivative of a function represents an infinitesimal change in the function with respect to whatever parameters it may have." The simple derivative of a function f with respect to x is denoted by $f'(x)$, which is the same as df/dx. Newton used fluxions notation $dz/dt = \dot{z}$, but it means the same, so I will use f' or df/dx from now on. Well, I can now take the derivative of certain classes of functions because I just follow certain rules.

If my function is of the type x^n, I use the fact that $d/dx(x^n) = nx^{n-1}$. So, if I have $f(x) = x^5$, its derivative should be $5x^4$. This is easy. If I analyze trigonometric functions such as $\sin x$ and $\cos x$, then I use the derivatives $d/dx(\sin x) = \cos x$, and $d/dx(\cos x) = -\sin x$.

Taking derivatives is so easy! I could spend hours deriving more complicated functions. However, I wish to learn also how to see the world through mathematics. I must find the connection between differential equations and physics. I am eager to explore this applied aspect of mathematics.

Let's start with a differential equation, an equation involving an unknown function and its derivatives. It can be relatively easy such as $\frac{dP}{dt} = kP$, or a bit more complicated such as the linear differential equation:

$$(x^2 + 1)\frac{dy}{dx} + 3xy = 6x,$$

or even a nonlinear equation such as this:

$$\frac{dy}{dx} = (1 - 2x)y^2.$$

A differential equation is linear if the unknown function and its derivatives appear to the power 1 (products of these are not allowed) and nonlinear otherwise. The variables and their derivatives must always appear as a simple first power. Nonlinear equations are difficult to solve and some are impossible.

First I need to master linear equations. Some mathematicians use the notation y' for the dy/dx derivative, or y'' for d^2y/dx^2, and so forth. Thus, the previous linear equation would be written as $(x^2 + 1)y' + 3xy = 6x$. I need to keep these differences of notation in mind, since I am studying from five different books.

I studied the properties of differential equations and learned to solve them. Now, I must learn how to apply differential equations. But how do I translate a physical phenomenon into a set of equations to describe it? It is impossible to depict nature in its totality, so one usually strives for a set of equations that describes the physical system approximately and adequately.

Say that I want to predict the growth of population in Paris. To do it, I can use an exponential model, that is, an equation that represents the rate of change of the population that is proportional to the existing population. If $P(t)$ represents the population change in time (t), I write $\frac{dP}{dt} = kP$, where the rate k is constant. I observe that if $k > 0$, the equation describes growth, and if $k < 0$, it models decay. The exponential equation is linear with a solution $P(t) = P_0 e^{kt}$, where P_0 is the initial population, i.e., $P(t = 0) = P_0$.

Mathematically, if $k > 0$, then the population grows and continues to expand to infinity. On the other hand, if $k < 0$, then the population will shrink and tend to 0. Clearly, the first case, $k > 0$, is not realistic. Population growth is eventually limited by some factor, like war or disease. When a population is far from its limits of expansion, it can grow exponentially. However, when nearing its limits, the population size can fluctuate. Well, I think that the equation I use to predict the rate of change of population can be modified to include these factors to obtain a result closer to reality.

Aristotle thought that nature could not be expected to follow precise mathematical rules. But Galileo argued against this point of view. He envisioned the experimental mathematical analysis of nature to be used to understand it. Newton was inspired by Galileo and later developed the laws of motion and universal gravitation. Newton, Leibniz, Euler, and other great people then created the mathematics that help us converse with the universe.

Oh, how glorious it is to speak such a language and understand the whispers from the heavens and the world around me.

Thursday — March 7, 1793

Someone said that mathematics is the queen of sciences. This must be true because, like art and music, mathematics stimulates the senses and lifts the spirit. My pursuit of mathematics has saved me from fretting and thinking about the social chaos and uncertainty of my country's future.

My mother began morning prayers with thanksgiving and ended with a plea for resolution. The economy is worsening and food is in short supply. Millie comes home from the market with fewer goods, as the shops are almost empty. In some parts of the city, hungry and desperate people attack butcher shops and bakeries to ensure their meager meals.

Abolishing the monarchy has not brought the changes that were supposed to benefit the citizens of the lower class. The ravages of the social revolution can be seen everywhere. There are many more hungry people, more homeless children. Poverty is rampant. Isn't this what the revolution was supposed to eradicate?

When darkness descends upon Paris, the false sense of security disappears. Soldiers patrol the streets; the sounds of guns going off mingle with the bells striking the hour. The Commune decree forbids people from walking after ten o'clock at night without an identity card. At night, Paris becomes a dungeon of terror and fear.

Sunday — March 10, 1793

It is a stormy night; from my window I see the misty and cold darkness illuminated from time to time by lightning. The bells of Notre-Dame have just struck ten and I sit here unable to sleep, thinking, worrying.

Friday — March 15, 1793

I mustered the courage to write a letter to Antoine LeBlanc. I asked whether he would permit me to study his lecture notes on mathematics. Mother would be mortified to learn that I was so bold. I must admit, I feel a bit anxious, afraid that Antoine will either ignore my letter or, worse, make fun of me and belittle my request. However, there is no turning back. He must have received my letter already.

After my illness, I resumed my study of linear differential equations. Thus far, I've learned that there are many types of equations with many levels of complexity. Mathematically, differential equations are beautiful and, if they are linear, the methods for solving them are straightforward. For example, a type of differential equation that I learned to solve is $\frac{dy}{dx} + p(x)y = q(x)$. It is a first order linear differential equation whose general solution is given by $y = \frac{\int u(x)q(x)dx + C}{u(x)}$, where $u(x) = e^{\int p(x)dx}$ is the integrating factor. I suppose I like these types of equations because they are easy to solve, and the solution involves Euler's number e. Differential equations can also be written in the form $y' + p(x)y = q(x)$, where $y' = dy/dx$.

But I discovered that there are complicated differential equations of first and higher order that are more difficult to solve. There are equations of the form $y' + p(x)y = q(x)y^n$, which cannot be solved using the approaches I've learned so far. I attempted several methods and obtained a solution, but I am not sure my result is correct. That's why I thought of writing Antoine. I enclosed my notes and asked him to show my analysis to his tutor.

For now I must stop obsessing over the letter. I wrote it formally and correctly, just as Maman taught me. If Antoine chooses to ignore it, I will forget about it and will find another way to learn advanced calculus. However, if he is friendly, Antoine LeBlanc can be my connection to sources of learning, including the scholars themselves. I must be patient.

Thursday — March 21, 1793

There is so much to learn that at times I feel overwhelmed. But when I get the solution to a challenging equation, I feel exhilarated and want to continue, to go much further. Especially when I find equations that relate to a physical problem I understand well.

I discovered that a phenomenon as simple as that of cooling objects could be represented by a differential equation. For example, when hot chocolate is poured into a cup, it immediately begins to cool off. The cooling process is rapid at first, and then it levels off. After a period of time, the temperature of the liquid chocolate reaches room temperature. Temperature variations for cooling objects follow a physical law that can be stated as "the rate at which a hot object cools is proportional to the temperature difference between the temperature of the hot object and the temperature of its surroundings."

I can write the cooling law with mathematics. With $T(t)$ representing the temperature of the hot object at time t, the rate of change of temperature as dT/dt, which is the derivative of the temperature with respect to time, and its relative temperature with respect to the ambient temperature as $(T - S)$, I write:

$$\frac{dT}{dt} = -k(T - S).$$

Here S is the temperature of the surrounding environment, and k is a positive constant of proportionality that depends on the physical characteristics of the object.

The above is a first order linear differential equation, quite easy to solve; it requires that one knows how to integrate. But first, I need a boundary condition, in this case, *an initial condition*. In other words, I need to know the temperature of the object at the beginning of the cooling process, at time $t = 0$. Then I write this initial condition as $T(t = 0) = T_0$.

To solve the differential equation, first I separate the variables and then integrate as follows:

$$\int_{T_0}^{T} \frac{dT}{T - S} = -k \int_0^t dt,$$

which gives after integration

$$\ln(T - S) - \ln(T_0 - S) = \ln\left(\frac{T - S}{T_0 - S}\right) = -kt.$$

This equation can also be written in the form

$$\frac{T(t) - S}{T_0 - S} = e^{-kt}.$$

The solution to the differential equation I developed is formally given by

$$T(t) = S + (T_0 - S)e^{-kt}.$$

Très bien. This solution was quite easy. However, I know other equations will be much more complicated to solve. But I will try!

Monday — *April 1, 1793*

Like dusk that separates light and darkness, an instant of time intangible yet real, I am here now, no longer a child, but on the threshold of my womanhood. Today I am seventeen. Now I cross a verge into my future. What is my *raison d'être*? What is the reason for my being? What is my purpose in life? I only know that I am eager for genuine knowledge and wish to give myself to the study of mathematics.

I tremble with anticipation at the future before me because, as in my dreams, mathematics will rule my world. I could no longer deprive my soul of the beauty and harmony of the mathematics that intermingles with my thoughts, like the notes in an exquisite symphony. I want to create mathematical ideas and to live in such a fascinating realm independent of the material world. I wish to uncover the mysteries that remain beneath unsolved equations. I wish to see nature through the eyes of mathematics.

In the last few years I've read about great scholars and learned so much from each one. But if I had to name one mathematician who motivated me the most I would have to choose between Archimedes and Euler. One of the greatest thinkers of antiquity, Archimedes inspired me, awakening my interest in mathematics. He taught me that one must be so consumed by mathematical passion that nothing else matters.

Euler, one of the greatest creators and discoverers of mathematics, who kindly, patiently, as he instructed a fairly ignorant princess, taught me new concepts, from the most fundamental principles of algebra and trigonometry to the ideas of infinity and calculus. Euler also taught me that rigorous analysis leads to elegant equations of extraordinary beauty, equations composed of the most simple and superb numbers found in nature. Euler revealed to me the exquisite equation $e^{i\pi} + 1 = 0$, an identity that embodies in its simplicity the most fundamental numbers that rule the mathematical universe.

But my learning is far from complete. I am fully aware that there is much ahead, a myriad of new mathematical truths that I must seek. For I know that I shall use mathematics to draw my vision of the world, the way an artist sketches it.

Adieu mon ami. I leave my childhood behind, and now I cross the threshold to embrace my womanhood. This is the beginning of a new chapter in my existence, for I now commence my life as a mathematician. I close the last page of my childhood diary with this theorem that has intrigued me for so long, and that I aspire to prove one day.

I shall call it Fermat's Last Theorem, because it was perhaps the last attempt of Pierre de Fermat to solve such a monumental problem. Fermat summarized his idea in this simple statement: The equation $x^n + y^n = z^n$ has no non-trivial positive integer solutions for $n > 2$.

By *trivial* it means $(0, 0, 0)$, $(0, 1, 1)$, and $(1, 0, 1)$, which are always solutions. Non-trivial solutions have no zeros so they can be characterized by $xyz \neq 0$. I could start with the method that Euler used to prove that

$x^3 + y^3 + z^3 = 0$ has no solutions, a method that seems successful for $n = 4$ as well. Yet, if Euler could not proceed further and show that this works for all $n > 4$, then perhaps this is not the correct approach. However, if Monsieur Fermat asserted that he had found the proof, and that the margin of his book was too small to contain it, then it means that the full analysis must be rather lengthy. My challenge then is to learn new methods to prove this theorem, *je cherche un défi à relever*!

6
Intellectual Discovery

Friday — April 19, 1793

Is there anything more fascinating than prime numbers? The basic notion of prime numbers is so simple that even as a child I could understand it. Prime numbers cannot be written as a product, except of themselves and 1. For example 17, one of my favorite numbers, can only be written as $17 = 17 \cdot 1 = 1 \cdot 17$, and that's all. However, prime numbers combine that beautiful simplicity with a mystifying nature and a profound meaning that only those who know and understand mathematics can appreciate. The prime numbers contain within them such mystery and intricacy that many problems remain unsolved after hundreds of years. Mathematicians conjecture about their nature; eventually some of those conjectures become theorems and are demonstrated, but other assertions remain unproved. For example, in 1742 Goldbach conjectured that every even number greater than 2 is the sum of two primes, but there is no proof of it and today this conjecture is unsolved.

Another intriguing characteristic of prime numbers is that although they seem so irregular as to appear to be random, one can also find a myriad of patterns just by arranging numbers in a certain way or by combining the prime numbers with some integers.

For example, the relation $n^2 + n + 1$ results in numbers that look like primes, but on close examination they are not. Let's substitute a few

187

integers:

$$n: \quad 1 \quad 2 \quad 3 \quad 4 \quad 5 \quad 6 \quad 7 \quad 8 \quad 9$$
$$n^2 + n + 1: \quad 3 \quad 7 \quad 13 \quad 21 \quad 31 \quad 43 \quad 57 \quad 73 \quad 91$$

When I see the sequence of primes less than $100 : 2, 3, 5, 7, 11, 13, 17, 19,$ $23, 29, 31, 37, 41, 43, 47, 53, 59, 61, 67, 71, 73, 79, 83, 89, 97$, I realize that there is a gap in my formula, because it fails to produce a prime when $n = 4, 7$, and 9. Thus I conclude that numbers produced by $n^2 + n + 1$ are not always prime.

This other formula $4n + 3$ also seems to produce prime numbers, because for $n = 1, 2$ it does, but then it fails! Now I understand how tempting it is to assume that a given formula produces primes until one verifies its validity.

The mixture of mystery and almost-randomness that characterize prime numbers is perhaps why I like them. Without even trying I'm compelled to seek patterns, and the more I find the more I search. It's like an addiction! Because there are not just two or three, I am sure there are infinite patterns one can find!

Examining a formula I ask whether there exist infinitely many primes of a certain form such as $n^2 + n + 1$, or $4n + 3$. Since the time of the ancient scholars it has been known that there are infinitely many primes — in fact Euclid provided such proof. And thus as a mathematician, I must attempt to prove that there are (or not) an infinite number of primes of a given form such as the examples above, or any other I may find in my research or that I conceive.

Monday — April 29, 1793

Prime numbers are part of *la théorie des nombres*, perhaps the oldest branch of pure mathematics. Its development started perhaps with Euclid who, hundreds of years ago, studied the properties of numbers and proved that the number of primes is infinite.

The next important work is that of Diophantus of Alexandria, the splendid book *Arithmetica* that deals with the solution of algebraic equations and provides a most comprehensive study on the theory of numbers. After Diophantus and up to the time of François Viète and Claude Gaspard Bachet in the early seventeenth century, mathematicians continued to study the nature of numbers, but without making new discoveries and little advance was made. It was not until Pierre de Fermat began research in perfect numbers and Diophantine arithmetic that new theorems were stated about prime numbers. Fermat's extensive work contributed more to accelerate the discovery of new concepts, and cleared new roads for mathematics. He produced a great number of theorems, but many remain without proof.

It seems that from the time of Fermat to that of Euler, mathematicians were devoted to the discovery or the application of the new calculus, and hardly anyone dealt with the theory of numbers. Euler was the first to revive this subject; the many articles which he published in the *Commentaires de Pétersbourg*, and in other works, prove how much he advanced the science of numbers, extending the initial foundation and adding many more layers of knowledge; this Euler did while at the same time making progress in many branches of pure and applied mathematics.

Euler's erudite research led him to prove two important theorems of Fermat. The first theorem: if a is a prime number, and x is any number nondivisible by a, the formula $x^{a-1} - 1$ is always divisible by a; and the second theorem: any prime number of the form $4n + 1$ is the sum of two squares.

Many other important discoveries related to prime numbers are included in Euler's *mémoirés*. One finds there the theory of divisors of the quantity $a^n \pm b^n$, the *Partitione numerorum*, which he also included in his 1748 book *Introductio in Analysin infinitorum*; the use of imaginary or irrational factors in the resolution of indeterminate equations; the general resolution of indeterminate equations of the second degree, by supposing that one knows a particular solution; the demonstration of many theorems on the powers of numbers and particularly of the conjectures of Fermat, that the sum or the difference of two cubes cannot be a cube, and that the sum or the difference of two biquadratics cannot be a square. Finally one finds in these same writings a great number of indeterminate problems solved by very clever analytical artifices.

I am determined to learn more about Fermat's theorems, and I especially would like to solve a proposition that states that any prime number $4n + 1$ is the sum of two squares, because for me it is as if he had said that the equation $A = y^2 + z^2$ is always resolvable as long as A is a prime number of the form $4n + 1$. One can add that the equation $A = y^2 + z^2$ will have but one solution.

From the above summary I've made a list of topics that I need to master before I attempt to prove the theorems of number theory.

Friday — May 3, 1793

Now I study indeterminate equations, those having more than one variable and an infinite number of solutions. For example, $2x = y$ is one of the simplest indeterminate equations, as is $5x^2 + 3y = 10$, which is quadratic. I can state that "an indeterminate equation is an equation for which there is an infinite set of solutions." Indeterminate equations *cannot* be directly solved from the given information.

I examine these expressions:

$$ax + by = c$$
$$x^2 - py^2 = 1$$

where a, b, c, and p are given integers (provided that p is not a square number), and I know they are indeterminate equations.

If a single equation involving two unknown numbers is given, and no other condition is imposed, the number of solutions of the equation is unlimited; for if one of the unknown numbers is assumed to have any value, a corresponding value of the other may be found. Such an equation is called indeterminate. Although the number of solutions of an indeterminate equation is unlimited, the values of the unknown numbers are confined to a particular range; this range may be further limited by requiring that the unknown numbers be positive integers.

In general, every indeterminate equation of the first degree, in which x and y are the unknown numbers, may be made to assume the form $ax \pm by = \pm c$, where a, b, and c are positive integers and have no common factors.

I begin to study this concept with a very easy indeterminate equation: $3x + 4y = 22$, where x and y are positive integers.

First I rewrite the equation,

$$3x = 22 - 4y$$
$$x = 7 - y + (1 - y)/3$$

or

$$x + y - 7 = (1 - y)/3.$$

Since the values of x and y are to be integral, $x + y - 7$ will be integral, and hence $(1-y)/3$ will be integral, though written in the form of a fraction. Let

$$\frac{(1 - y)}{3} = m,$$

an integer. Then

$$1 - y = 3m$$
$$y = 1 - 3m.$$

I substitute this value of y into the original equation:

$$3x + 4(1 - 3m) = 22$$
$$x = 6 + 4m.$$

The equation $y = 1 - 3m$ shows that m may be 0, or may have any negative integral value, but *cannot* have a positive integral value.

The equation $x = 6 + 4m$ further shows that m may be 0, but cannot have a negative integral value greater than 1. So, m may be 0 or -1.

Therefore,

$$\left. \begin{aligned} x &= 6 \\ y &= 1 \end{aligned} \right\} \quad \text{or} \quad \left. \begin{aligned} x &= 2 \\ y &= 4 \end{aligned} \right\}.$$

Voilà! That was rather easy.

Now I solve a different problem, which is stated as follows: *Find the least number that when divided by 14 and 5 will give remainders 1 and 3, respectively.*

First, I represent the number by N, so I form the following relationships:

$$\frac{N-1}{14} = x, \quad \text{and} \quad \frac{N-3}{5} = y$$
$$N = 14x + 1, \quad \text{and} \quad N = 5y + 3$$
$$14x + 1 = 5y + 3$$
$$5y = 14x - 2$$
$$5y = 15x - 2 - x$$
$$y = 3x - (2 + x)/5$$

Now let $\frac{2+x}{5} = m$, an integer

$$x = 5m - 2$$
$$y = \frac{1}{5}(14x - 2), \quad \text{from the original equation}$$
$$y = 14m - 6.$$

If $m = 1$, $x = 5$, and $y = 8$. Thus, my answer is,

$$N = 14x + 1 = 5y + 3 = 43.$$

Très bien. These indeterminate equations are easy. I need a greater challenge. It is time for me to attempt more advanced mathematical methods and be better prepared to prove more complicated theorems.

Thursday — May 23, 1793

The queen. It seems that people have forgotten about her or at least her name is no longer mentioned in the newspapers. But I often wonder about Marie-Antoinette and her children, still imprisoned in the Temple. I imagine how she must feel, mourning her husband, killed at the hands of those who were once the subjects in his kingdom. How do the children pass the time in that frightful dungeon? Does the little one play and amuse himself as other young boys do? What about the princess? When I see my sister Angélique I think of Marie-Thérèse who was her same age, 14, when they were thrown in that awful prison. Are they being tormented by their captors? Nobody is there to defend them. Oh, poor creatures, deprived of the most basic comforts, the little ones deprived of their innocence. Why are some humans so cruel sometimes, using the excuse of patriotism?

Friday — May 31, 1793

I had to go out. Not that I dislike my seclusion or crave the bustle of the city. No. I needed to feel the warmth of the summer sun burning my pale skin, to cleanse my emotional wounds and soothe my fears. I needed to take a stroll by the *quai de Louvre*, to hear the water flowing under the Pont-Neuf, and just to feel as I did when I was a child.

My mother would not hear of it. Worried about the dangers that lurk in every alleyway, first she refused to let me go. She is fearful of the drunkard sans-culottes who are mad with misguided fervor and accost women in the streets. But in the end Maman relented, after I promised her that I would not go near the Conciergerie. Millie came with me, eager to leave the house on an errand more interesting than simply going to market or waiting in her queue outside the bakery. They both forced me to wear the tri-colored ribbon on my hat, as there is a rumor that the Convention will soon make it mandatory that all women wear the patriotic insignia in public.

We walked all the way, as M. Morrell had left earlier with the carriage to fetch Papa and no cabs were to be found by Saint-Denis. The weather was hot but not as oppressively blistering as it had been days earlier; it is rather surprising to find so many people walking about, carefree and seemingly undisturbed by the daily news of arrests and executions.

On a whim I chose to turn right on rue Saint-Honoré, which as always was full of shoppers. We then turned left on rue du Pont-Neuf, not really planning to cross the bridge. But I'm glad we did because, upon stepping out into *l'Ile de la cité*, I discovered a bookshop specialized in mathematics, right on the corner of the *quai des Augustins*. The shop was empty when we entered, thus allowing me to browse without curious eyes and questioning stares. What I found is indescribable, a treasure of knowledge neatly recorded in the pages of those volumes. New books written by contemporary mathematicians and exquisite translations of ancient manuscripts, all bound in beautiful leather covers and inscribed with golden letters. I was glad that I brought a full purse and purchased *Oeuvres d'Archimède* for 30 francs. It is expensive but I couldn't let go of it. The book begins with a biographical portrait of Archimedes, then it discussed his memoirs, and it is supplemented with a history of the arithmetic development in Greece.

I don't remember the walk back home, as I was anxious to return and read my treasure.

Tuesday — June 11, 1793

There are different methods to prove theorems. One method is "mathematical induction," used for proving a statement — a theorem, or a formula — that is asserted about every natural number. By every natural number, or all natural numbers, I mean any number that I might possibly name.

Mathematicians use induction as a way of formalizing a proof so that one doesn't have to say "and so on" or "we keep on with a similar argument" or some such statement. The main idea is to show that the result is true for $n = 1$ and then demonstrate how once one has shown it to be true for some integer, we can see that it must be true for the next one as well.

These are the two steps of mathematical induction:

If (I) when a statement is true for a natural number $n = k$, then it will also be true for its successor, $n = k + 1$; and

(II) the statement is true for $n = 1$; then the statement will be true for every natural number n.

For, when the statement is true for $n = 1$, then according to (I), it will also be true for $n = 2$. But that implies it will be true for $n = 3$, for $n = 4$, and so on. In other words, it will be true for any natural number that I may consider. To prove a statement by induction, then, I must prove parts (I) and (II) above. Let's begin with a simple example.

Theorem. *The sum of the first n odd numbers is equal to the nth square.*

That means that $1 + 3 + 5 + 7 + \cdots + (2n - 1) = n^2$.

Proof.

a) First I make the following assumption: The statement is true for $n = k$:

$$1 + 3 + 5 + 7 + \cdots + (2k - 1) = k^2.$$

b) On the basis of this assumption, I show that the statement is true for its successor, $k + 1$:

$$1 + 3 + 5 + 7 + \cdots + (2k - 1) + 2k + 1 = (k + 1)^2.$$

c) Now I show it by adding $2k + 1$ to both sides of my induction assumption:

$$1 + 3 + 5 + 7 + \cdots + (2k - 1) + 2k + 1 = k^2 + 2k + 1$$
$$= (k + 1)^2.$$

d) To complete the proof by induction, I must show that the statement is true for $n = 1$.

e) So I show that: $1 = 1^2$.

I have now fulfilled both conditions of the principle of mathematical induction. The theorem is therefore true. Q.E.D.

Since the proof was easy enough I wonder if I should use mathematical induction to prove that $4n + 1$ is a prime number.

Wednesday — June 19, 1793

I saw him. Antoine-August was at the corner of Saint-Denis and rue au Fers, with a thick stack of papers clenched in his left hand and an austere frown marking his forehead. We exchanged glances and went our way. Millie recognized him first, for she elbowed me and whispered some silly thing I did not understand. I was not sure if I should be the one to acknowledge his presence or wait for him to wave his hand or speak to me. Antoine hurried across towards rue Saint-Honoré and disappeared.

I wonder now whether my last letter did not reach him, or maybe he got tired of the exchange. Last time I wrote begging him to help me find a book that contains Euler's totient theorem and also asked to show his tutor my proof of a little theorem, just to assure me if it is right. Well, I don't care that he didn't speak to me. It doesn't matter now if Antoine does not respond to my letters, as I have found the book I was looking for. And I am certain my proof is correct.

Sunday — July 7, 1793

What is the sum of all positive integers from 1 through n? Clearly, if $n = 1$, the sum equals 1, if $n = 2$, the sum is 3, and so on. But a better question is whether there is a method or formula that would allow me to determine the sum for any value of n without having to add the numbers or summands. I start with $n = 5$, for example, and determine the sum of the first five consecutive positive integers, i.e., $1 + 2 + 3 + 4 + 5 = 15$. From there I notice that in each case the sum, which I will call S_n, equals one-half the product of n and the next integer, $S_n = 1 + 2 + 3 + 4 + 5 = (5)(6)/2$; in other words,

$$S_n = \frac{n(n+1)}{2}, \quad \text{for } n = 1, 2, 3, 4, 5.$$

Now, if I want to make sure that this formula would work for any n greater than 5, I'd write a theorem that states:

"The sum of all positive integers from 1 through n is $S_n = \frac{n(n+1)}{2}$, for all $n \geq N$."

To prove by induction that for all n, the sum is $S_n = \frac{n(n+1)}{2}$, I begin by establishing that the sum is true for $n = 1$, so I write $1 = \frac{1(1+1)}{2}$.

Now I assume that the statement is true for an arbitrary number $n = p > N = 1$:

$$S_n = 1 + 2 + 3 + \cdots + p = \frac{p(p+1)}{2}.$$

Under the induction hypothesis, the theorem must then also be true for $n = p + 1 = q$:

$$S_{n+1} = 1 + 2 + 3 + \cdots + p + (p+1) = \frac{p(p+1)}{2} + (p+1)$$

$$= \frac{p(p+1)}{2} + \frac{2(p+1)}{2} = \frac{(p+1)(p+2)}{2} = \frac{q(q+1)}{2}$$

I know that the formula is true for $n = 1$: choosing $p = 1$, it will be true also for $n = 1 + 1 = 2$, $n = 2 + 1$, etc; that is, the formula is true for all integers $n \geq 1 = N$. Q.E.D.

Monday — July 8, 1793

Mon Dieu! Mon Dieu! A warrant was issued to arrest the Marquis de Condorcet! He has been condemned and declared to be *hors la loi*, an outlaw!

Papa just gave me the startling news. I was shocked but he didn't seem surprised. He thinks that since M. Condorcet is opposed to the death penalty and argued strongly against the execution of King Louis XVI, the radical group leaders were against him. Not long ago Condorcet drafted a constitution for the new republic, a document that was considered too liberal; some people called it the Girondin Constitution even if M. Condorcet is not a member of the Girondins. However, since May of this year the Girondists have fallen from favor and the Jacobins, a most radical political group led by M. Robespierre, are leading the government. Foolishly defending his own draft M. Condorcet spoke openly against the one proposed by the Jacobins. This act alone led to his condemnation.

I don't understand it. The Marquis de Condorcet, an esteemed philosopher and mathematician, a learned and enlightened man of great integrity, has advocated so many excellent reforms for France, such as the education of women and all children, and the establishment of a self-regulating educational system under the control of a National Society of Sciences and Arts to protect education from political pressures.

However, Papa stated that the Legislative Assembly has been hostile to all autonomous corporate structures and thus ignored Condorcet's plan. The Jacobin-dominated National Convention considered it too moderate. Then Marie-Jean Hérault de Seychelles, a member of the Constitution's Commission, misrepresented many ideas from Condorcet's draft and presented his own document, the same one Condorcet criticized in a letter published in a newspaper. Chabot denounced the letter to the Convention in today's session. His enemies used this act to brand Condorcet a traitor. And immediately the government issued a warrant for his arrest.

I am so afraid for M. Condorcet. Yet, I'd like to think that his many friends and colleagues will come forward and support him. They must convince the members of the Convention that he is not a traitor. Someone must speak up and remind the Tribunal that Condorcet is a leader of the Republican cause that can help make France a better nation.

When he was member of the Academy of Sciences, M. Condorcet produced a number of important works and helped to popularize science. In 1772, Condorcet published a work on the integral calculus which was described by Lagrange as "filled with sublime and fruitful ideas which could have furnished material for several works." Condorcet also studied the philosophy of mathematics and probability. His *Essay on the Application of Analysis to the Probability of Majority Decisions*, published in 1785, deals with the theory of probability. Papa has high esteem for M. Condorcet perhaps because his work favors Turgot's economic theories and agrees with Voltaire's opposition to the Church.

For me, Monsieur Condorcet is especially important, because he gave me a gift of mathematics (Euler's book *Introductio in analysin infinitorum*) that I will treasure forever. If I could, I would defend this great man who himself has defended the human rights of the oppressed and victimized, a man of lofty principles who speaks to abolish slavery and promotes the education of all women. All I can do for M. Condorcet is wish for the best.

Saturday — July 13, 1793

It is past midnight and I cannot sleep; my mind is troubled by the disturbing news delivered by Monsieur LeBlanc a little while ago. He came unannounced and out of breath to say that a young woman has just assassinated M. Jean-Paul Marat. At 7:30 this evening Citoyen Marat, the leader of the Jacobins and a radical journalist, was confined to his bathtub when a lady named Charlotte Corday, a Girondin supporter from Normandy, stabbed him to death! She had come earlier and finally was admitted to Marat's room in his house on the pretext that she wished to claim his protection.

Nobody knows why she killed Marat, but what is known is that Mlle Corday is an aristocrat who sympathizes with the aims of the Revolution. She didn't flee but instead waited calmly to be arrested. Shortly after, officials of the Committee of General Security arrived at the scene and found in her possession a copy of Plutarch's *Parallel Lives*, a book of biographies of Greek and Roman leaders. Clearly she is a well-educated woman and must have been horrified by the violent acts of the Jacobins. Mlle Corday is now in the prison de l'Abbaye, the same dungeon where last year a hideous massacre took place.

My father and his friend speculated about this woman and her plan for killing *l'Ami du people*, as everybody calls Marat. They reflected on her motives, thinking that they were based in part on tyrannicide, which is an act of horrible desperation. Though the Catholic doctrine condemns tyrannicide, great theologians of the Church like St. Thomas have accepted rebellion against oppressive rulers when the tyranny becomes extreme and when no other means of safety are available. In other words, the assumption is that Mlle Corday must have believed that she would stop the murder of thousands by executing Marat. I wanted to hear no more and came to my room to calm down.

I could not use this kind of reasoning to justify murder. It's just so difficult to believe that a woman would resort to taking the life of another human being, even if she felt warranted. She is so young, not even 25 — I wonder whether the people of Paris will rise in admiration or rage. There is no question she will be condemned to the guillotine. Charlotte Corday committed a violent crime against one of their own, the jury will have no mercy.

Marat was a madman who signed execution orders without mercy, an ardent patriot and paradoxical man who fiercely persecuted the "enemies of the people." Today the friend of the people is no more.

Thursday — August 1, 1793

More disturbing news arrived today. In the middle of the night the deposed queen was taken to the Conciergerie. Millie came from the Halles at 11 this morning eager to recount the gossip she heard from the market women. My mother calls the Conciergerie "the antechamber of death" because anyone who is incarcerated there will end up at the guillotine. Maman could not contain her tears when Millie told us that sometime around two o'clock in the morning the queen was roused from her bed by some Commissioners of the Commune and was taken to the Conciergerie. She was not even allowed to say good-bye to her daughter the young princess or to Madame Elisabeth, and now she is in the worst cell of the prison.

How can the new republic build its foundation upon the suffering and torment of women? Even Marie-Antoinette does not deserve this inhuman treatment. Maman has cried since July when the Convention announced that the young prince Louis would be taken from his mother; the poor child is only eight! I am sure that after losing her little boy nothing must matter to this wretched woman...

Thursday — August 8, 1793

The Academy of Sciences was closed today. The Convention Nationale has decreed the abolition of all academies and learned societies licensed or endowed by the Nation. How is it possible that the most important institution for science and a center for scholars to advance knowledge be considered a threat? But as Papa says, being a royal institution the Academy is an obvious target of the revolutionaries and, although some argued that it serves a useful purpose for the nation and should be exempted, even some of its members accepted that it was undemocratic and approved its abolition.

The extensive and precious scientific collections of the learned society will be placed under lock and key. What about the men of science who are the foundation and structure of this institution? *Laplace, Legendre, Monge et Lagrange, estimables savants, c'est par vous que l'univers devra ce bienfait à la France.*

But my father reassured me that the philosophers and mathematicians will continue to assemble somewhere else and will continue discussing their scientific endeavors. He believes the savants will meet to address issues that will be commissioned by the Convention. Of course, no nation can live without scholars. But a scholar doesn't need to engage in a society or seek membership in an organization to advance and develop new science or to work in the interest of human knowledge. A scholar can work in the silence of his own study, just like me.

Saturday — August 17, 1793

For more than a year I have been making a list of prime numbers with the intention of having a look-up table to check against if a given integer is (or not) a prime. It was a tedious process, verifying whether a number is prime, one by one, but some days I just couldn't stop, curious to see how many more I'd discover. Today, upon confirming that 3571 is prime, I realized that my table already contains the first 500 primes!

Of course I know the number of primes is infinite, but they seem to occur less frequently as one goes farther out in the number sequence. Since the time of Euclid, mathematicians have tried to formulate a law relating the number of primes less than, or equal to, a given integer n. Perhaps one of the ultimate goals should be to discover a law that establishes the number of primes all the way to infinite. That would be sublime!

Saturday — August 24, 1793

Fifteen shots! Yes, I counted. Fifteen rounds of ammunition were fired at 20 past eleven. The frantic voices pierced the darkness of the night while the thunder of cannon echoed in the distance. I was startled but no longer surprised, since I've become oblivious to this nocturnal chaos.

Yesterday the Convention decreed the *levée en masse*, mobilizing the entire nation. Carnot exhorted: "The young men shall go to battle; the married men shall forge arms and transport provisions; the women shall make tents and clothing and shall serve in the hospitals; the children shall turn old linen into lint; the aged shall betake themselves to the public places in order to arouse the courage of the warriors and preach the hatred of kings and the unity of the Republic." The intent of the levy is to take control of an unruly society that requires discipline if it hopes for the Republic to be victorious. Now, however, virtual anarchy rages on the streets beneath my window.

I wait now for the quietness to resume my reading. When I study and perform my analysis the horrors of the revolution seem far away and cannot touch me.

Tuesday — September 3, 1793

Some time ago, when I was looking for a formula to generate prime numbers, I read about Marin Mersenne and his conjectures. He thought that $2^n - 1$ is prime for $n = 2, 3, 5, 7, 13, 17, 19, 31, 67, 127,$ and 257, and composite for all other positive integers $n < 257$. Well, I just learned that in the year 1536, a mathematician named Hudalricus Regius had already shown that $2^{11} - 1 = 2047$ was not prime. Amazing! That was the same number I used to demonstrate that Mersenne's assertion was flawed.

Also, in 1603 the Italian mathematician Pietro Cataldi correctly verified that $2^{17} - 1$ and $2^{19} - 1$ are both prime, but he incorrectly stated $2^n - 1$ is prime for 23, 29, 31 and 37. Then in 1640 Fermat showed Cataldi was wrong about 23 and 37. Almost a hundred years later Euler showed Cataldi was also wrong about 29. In further studies, Euler admitted that Cataldi's assertion about 31 was correct. I am not sure when Euler proved this, because I don't have copies of all his memoirs, but from my research I deduce it was in October 1752 when Euler wrote a letter to Goldbach stating that he was uncertain about this number (even though he had earlier listed it as prime). Finally, in 1772 Euler wrote to Bernoulli stating that he had proved $2^{31} - 1$ prime by showing all prime divisors of $2^{31} - 1$ must have one of the two forms $248n + 1$ and $248n + 63$, and then dividing by all such primes less than 46339. Euler's result must have required the proof of a theorem that I am still trying to find.

Now I study a different topic, stimulated by a memoir Euler wrote in 1732 with the title *Observationes de theoremate quodam Fermatiano aliisque ad numeros primos spectantibus* (Observations on a theorem of Fermat and others regarding prime numbers). According to Euler, Fermat asserted that any number of the form $a^n - 1$ is a prime number. But Euler proved that it is not!

On the third page, Euler wrote: "whenever n is not a prime number, not only $2^n - 1$ but also $a^n - 1$ has divisors. But if n is a prime number still $2^n - 1$ can be seen to exhibit one, although no one has ventured to assert this insofar as I am aware, since it could easily have been refuted." Euler goes on to say that $2^9 - 1$ is not a prime number because it is divisible by $2^3 - 1$, i.e., 7.

Euler also stated that $2^{n-1}(2^n - 1)$ gives a perfect number whenever $2^n - 1$ is prime and therefore n ought to be a prime number. He asserted that if $n = 4m - 1$ and $8m - 1$ is a prime number, then $2^n - 1$ can always be divided by $8m - 1$. And because $11, 23, 83, 131, 179, 191$, etc. are excluded from the formula, it makes $2^n - 1$ a composite number.

At the end of his memoir, Euler added six elegant theorems without proof. He wrote:

"Haec persecutus in multa alia incidi theoremata non minus elegantia, quae eo magis aestimanda esse puto, quod vel demonstrari prorsus nequeant, vel ex eiusmodi propositionibus sequantur, quae demonstrari non possunt, primaria igitur hic adiungere visum est."

My translation from Latin is not that good, but I think Euler admits that he did not yet know how to prove them. Or maybe Euler left those theorems for someone like me to demonstrate.

Tuesday — September 17, 1793

Madness has spread through my beloved France. The National Convention decreed the Law of Suspects, which authorized the imprisonment of almost anyone. My father said that the intolerant sans-culottes once again invaded the Convention to exert pressure on the deputies. They wanted to impose economic measures to ensure their food supplies, and asked the government to deal more harshly with the counter-revolutionaries. A delegation of the forty-eight sections of sans-culottes urged the Convention to "make Terror the order of the day"!

On the fifth of September, the Convention voted to implement terror measures to repress *counter-revolutionary* activities. In simple words, a man (and his family) might go to the guillotine for saying something critical of the revolutionary government. If an informer happens to overhear, that is all the tribunal needs to convict a person. Watch Committees around the nation were encouraged to arrest "suspected persons, those who, either by their conduct or their relationships, by their remarks or by their writing, are shown to be partisans of tyranny and federalism and enemies of liberty."

As of today, my father's weekly meetings with his colleagues are cancelled. They all fear being accused of something, even if they are supportive of the Republic. The slightest hint of counter-revolutionary opinions or activities can place anyone under suspicion. It is a hard time that we all must endure, Papa says. And he asked my mother especially to be strong and not give herself over to undue anguish. But I know my father keeps from her the dread he must feel at the prospect of being accused. Nobody is safe now.

Monday — September 23, 1793

To pass the time my mother works on tapestry and knits mittens for the poor orphan children. Although she no longer directs my studies, my mother asks me to interpret passages of our favorite books. She is frequently at prayer, and every morning reads the appointed service of the mass for the day; she also studies her religious books, and sometimes, at our request, she recites a poem or reads some pages of an interesting play. On Saturday, when we assemble after supper, Maman sings her arias accompanied by Angélique at the piano. My sister also plays little interludes and enchantingly gentle sonatas. I simply listen and daydream or I think about numbers.

Thursday — October 3, 1793

The name of Condorcet reappeared today in the list of the conventionals brought before the revolutionary tribunal, accused of conspiracy against the unity of the republic. After such hearings these people are condemned to death. But Papa says M. Condorcet was not among the deputies arrested and that he has been hiding since July. Nobody seems to know where he is, and thus his name is also in the list of exiles.

Madame Condorcet, his wife, is now penniless because the government confiscated all his possessions. Papa assured me that she is doing fine and is now living in a small apartment not far from here in rue Saint-Honoré. Now Madame Condorcet has to support herself and provide for their four-year old daughter and her younger sister. Although she is educated, she was obliged to open a shop to survive financially. She is also an artist so she paints portraits that she sells very well.

Ironically, the warrant for her husband's arrest is in full view of lugubrious notices, near her store, declaring in large characters that the punishment of death would be inflicted upon anyone who rendered assistance to those who have eluded the law. Nobody knows where M. Condorcet is now. Has he escaped to a foreign land? It is unlikely, as he would have been caught at the border without a passport. Papa thinks that he could be in Paris, hiding in some friend's attic. I hope so.

Friday — October 11, 1793

Here is an interesting statement: "There is always one prime between n^2 and $(n+1)^2$ for every integer greater than zero." This assertion I write as follows:

$$n^2 < p < (n+1)^2, \text{for } n > 0.$$

If I call N the number of primes p between n^2 and $(n+1)^2$,

n	1	2	3	4	5	6	7	8	9	10
N	2	2	2	3	2	4	3	4	3	5

I see that the minimum value of N is always 2. Never 1! I've checked this result for many values of n greater than 10 and find that N oscillates but shows a general upward trend as n increases. Thus, if indeed N is always 2 or greater, would it be better to say that "for every positive integer n there exist at least 2 primes p and q such that $n^2 < p < q < (n+1)^2$, for $n > 0$"?

Monday — October 14, 1793

The trial of Marie-Antoinette has concluded; her defense lawyers were unable to save her. The widow Capet, as many call her, was accused of betraying her country. Papa says that her noble attitude, even in the face of the atrocious accusations of Fouquier-Tinville, commanded the admiration even of her enemies, and her answers during her long examination were clear and skillful. Yet, the jury decided unanimously to convict her — without proof! But then again, everybody knew how the trail would end. Will they let her go? Papa thinks not.

Wednesday — October 16, 1793

Today, at fifteen past twelve, time stood still, at the moment when France used a horrible way to annihilate its monarchy. We are mourning a woman of great courage, Marie-Antoinette, our former queen. At home, before the wretched hour we assembled in silence to grieve together what none of us could prevent.

At five o'clock in the morning we heard the *rappel* beaten in all the Sections, and by seven o'clock the armed force designed to guard the streets between the prison and the scaffold was ready. At ten o'clock the turnkey was sent by the concierge into the cell of the queen. I was not there to see it but it is not difficult to imagine what happened next.

What tormented nights must Marie-Antoinette have spent in her wretched cell, with the prospect of a terrible death always in her mind! Maman said several times that the queen must have suffered perhaps more just thinking of her children, not knowing into what hands the dauphin had fallen or what her captors would do to her precious daughter, still in the dark prison with her young aunt, with no one else to comfort them, knowing all too well her mother would be executed. The queen spent seventy-five days in the Conciergerie, taken there on the night of the 2nd of August, and leaving it this morning to be executed at the guillotine. Marie-Antoinette was but thirty-eight.

Why? She was in no way dangerous to the Republic, and even the leaders of the government hesitated for weeks to take her life. The king has been dead for months. They have all they could want, and they have destroyed the direct line of their monarchs. I do not understand it! They could have let the former queen go to Austria, if she had asked, or simply let her live her last days somewhere in France. What could she do now without the authority of the monarchy or without a powerful family and friends?

I did not witness her wretched ride down rue Saint-Honoré. But Millie, who saw her was awed; she told us the queen looked dignified, stoic, and courageous with her hair cropped and her hands tied behind her back. I didn't want to hear Millie's words yet I remained transfixed listening to her excited account of the tragic event.

The queen was dressed in a white peignoir, and wore a *fichu de mousseline* crossed over her breast. On her head was a little plain white linen cap. On this morning, when she was about to rejoin her husband, it is clear the former queen would wear no mourning. A tumbril, drawn by a white horse waited outside the Conciergerie. Marie-Antoinette mounted the cart with difficulty, owing to her bound hands. Millie noted she seemed calm and oblivious to the cruel cries of the mob yelling insults and tormenting her more. I don't know if I should believe it although Millie swears she smiled at the queen and made the sign of the cross to give her comfort, and that the lady seemed grateful.

I do not wish to imagine what must have been the last sad moments of the former queen of France. No human being deserves such a gruesome death.

Thursday — October 24, 1793

Today is no longer Thursday, October 24 of the year 1793. According to our new calendar, today is "3 Brumaire year II." It is hard to believe but Papa just brought home a copy of the calendar that France has adopted. The Republican government decided to reorganize the system we have used for centuries to measure time and record history. The National Convention, after having heard the recommendation of its Committee of Public Instruction (composed of astronomers, mathematicians, poets, and dramatists), decreed that all public acts shall be dated following the new almanac.

The calendar was designed to make a complete break with both the French aristocracy and their institutions, and also to sever ties with the Catholic Church, including the Gregorian calendar. Against the superstition and fanaticism, Sunday and the Christian festivals are abolished in the name of reason, science, nature, poetry, and other non-religious ideas.

Under the new calendar, a year consists of 365 or 366 days, divided into 12 months of 30 days each, followed by 5 or 6 additional days or "*jours complémentaires*," and leap years are to have an extra *jour complémentaire*. The poets on the committee chose the names of the months: Vendémiaire (September), Brumaire (October), Frimaire (November), Nivose (December), Pluviôse (January), Ventose (February), Germinal (March), Floréal (April), Prairial (May), Messidor (June), Thermidor (July), and Fructidor (August). The calendar names the days after an animal (days ending in 5), a tool (days ending in 0) or else a plant or mineral (all other days); there are 366 new day names! How am I going to remember them?

Well, that is fine to change the names of the days and months, but the trouble with this calendar is that the months are not divided into weeks; instead each month is divided into three décades of 10 days, of which the final day is a day of rest. I am sure this will create problems because now there are 9 work days between each day of rest, whereas before we had only 6 work days between each Sunday. The ten days of each décade are called, respectively, Primidi, Duodi, Tridi, Quartidi, Quintidi, Sextidi, Septidi, Octidi, Nonidi, and Decadi.

Now years will be counted starting with the establishment of the first French Republic on 22 September 1792. That is why today we are already on the second year (a. II). I will have to have a copy of the new calendar by my desk to remind myself what day it is!

Sunday — November 3, 1793

I was thinking about prime numbers of the form $4n+3$, and wonder whether I could use Euclid's method to prove that there is an infinite number of primes of this form. In other words, could I prove that the set of all primes of the form $4n + 3$ is infinite, by using as a model Euclid's proof that there are an infinite number of primes?

This idea came to me because I know that the product of several numbers of the form $4n + 1$ is always of the same form, and the product of an odd number of numbers of the form $4n + 3$ is also of the form $4n + 3$; but the product of an even number of numbers of the form $4n + 3$ is of the form $4n + 1$.

Therefore, if I multiply, say, the first six terms in the $4n + 3$ sequence,

$$3 \cdot 7 \cdot 11 \cdot 19 \cdot 23 \cdot 31$$

and add 2,

$$3 \cdot 7 \cdot 11 \cdot 19 \cdot 23 \cdot 31 + 2$$

I get a number of the form $4n + 3$ that is not divisible by $3, 7, 11, 19, 23$, or 31. Its prime factors cannot all be of the form $4n + 1$, so at least one factor must be a "new" prime of the form $4n + 3$. If I call this new prime p, and could form a new expression. Or maybe I could prove my original assertion using indirect proof by contradiction.

Sunday — November 10, 1793 (20 Brumaire, année II)

I must write down what happened today, but I am not sure if I can express well my emotions. I can only record what I saw and try to describe an event in which I took no part; even if my body was there, I wasn't!

The first *Fête de la Raison* was held in the cathedral Notre-Dame, now deconsecrated and renamed the "temple of reason." This was an event meant to emphasize the secular principles of knowledge, reason, and political liberty.

We didn't want to go, especially my mother, but in the end she relented, thinking that refusing to go would only cast a suspicious light on our family, especially on Papa. Upon arrival, Maman made no effort to conceal her horror when she discovered that the statues on the church's façade were beheaded. Those who performed such a senseless act must have believed that the statues were of French kings when in fact they were Old Testament kings. How ignorant! Inside the church was no better; its sacred treasures were looted and destroyed.

But oh, the most outrageous act was the drama that unfolded inside; the Festival of Reason was nothing but a theatrical play with a woman taking the role of Liberty. She was wearing a white dress, a long blue cloak,

First *Fête de la Raison* (Festival of Reason) at Notre-Dame — Paris, November 10, 1793.

and a *bonnet rouge*. Many young girls dressed in white and wreathed with oak leaves, singing patriotic hymns, preceded lady Liberty into the flower-decked "temple." Four men carried the "goddess of Liberty" on her throne up the aisle towards a flame representing "the torch of truth" that burned alongside busts of revered philosophers. Hanging above the altar there was a banner with embroidery depicting a tree of liberty, equality, and reason, spreading its roots over the world. Beneath the tree were scattered a torn-up Bible and fragments of crowns and scepters, the remnants of religion and royalty.

The Commune's president, M. Chaumette, gave a speech decrying fanaticism. People responded with *"Vive la Montagne!"*, *"Vive la République!"* and *"Vive la liberté!"* But for me liberty, reason and truth are only abstract concepts. These are not gods, or goddesses for that matter, and properly speaking reason and truth are parts of ourselves. Why, if the purpose of the Republic is dechristianization and to put an end to idolatry, why did the government use this grotesque display to indoctrinate us with a woman taking the role of a goddess?

Tuesday — November 19, 1793

My poor mother is overcome by a melancholy so terrible that it scares me. She sits whole hours in silent despair; and her only consolation is her grandchild. She waits every Saturday for Madeleine to come visit her with baby Jacques-Amant. He is already two years old. The only pleasure my mother enjoys is seeing him uttering his first words and seeing him toddling around. She watches through the window, waiting for the child to show up in the arms of my sister; those weekly visits are her only hope, her only thought, and her anchor to life itself.

Angélique is the only one who persuades Maman to talk, asking for fashion advice and even makes her laugh with her silly chatter. My sister daydreams and then makes up childish stories. The hero in her tales is always "a handsome gentleman who will come from a distant land to rescue her;" he is a young man who will take us all to a country where people are not afraid and live in peace. I am not sure there is such an idyllic Eden.

Saturday — December 7, 1793

How many grains of sand would fit in the "universe"? One thousand myriads of eight numbers, according to Archimedes! He developed a number system that uses powers of a myriad capable of giving numbers as large as the number of grains in the universe. "Universe" — Archimedes said — is the sphere whose center is the center of the earth and whose radius is equal to the straight line between the center of the sun and the center of the earth (compared to the sphere of stars).

Big numbers had a special fascination for Archimedes. Among his many contributions to mathematics, *The Sand Reckoner* is probably his most popular book. With this work Archimedes introduced a new system for generating and expressing very large numbers. He must have devised this in frustration, as very large numbers were clumsy to express in ordinary notation at that time.

The Greeks used the twenty-seven letters of their alphabet for numerals (except that they didn't have a sign for zero) and their mathematical notation was not positional; it utilized many symbols and, for a mathematician, it was rather cumbersome. Archimedes contrived a procedure involving indices to reduce the problem to more manageable dimensions. For the number 10,000, a *myriad*, the letter M was used. The Greek word is *murious* (uncountable, pl. *murioi*). The Romans converted this to *myriad*. The M could be combined with other letters and signs to make bigger numbers.

In *The Sand Reckoner* Archimedes argued that his procedure gives numbers large enough to count the number of grains of sand which could fill into the universe. He wrote:

> "There are some, King Gelon, who think that the number of the grains of sand is infinite; and I mean by the sand not only that which exists about Syracuse and the rest of Sicily but also that which is found in every other region whether inhabited or uninhabited.
>
> Aristarchus of Samos brought out a book consisting of some hypotheses, in which the premises lead to the result that the universe is many times greater than that now so called. His hypotheses are that the fixed stars and the sun remain unmoved, that the earth revolves about the sun in the circumference of a circle, the sun lying in the middle of the orbit, and that the sphere of the fixed stars, situated about the same center as the sun, is so great that the circle in which he supposes the earth to revolve bears such a proportion to the distance of the fixed stars as the center of the sphere bears to its surface."

Archimedes was referring to the Greek mathematician and astronomer Aristarchus of Samos (c. 310 B.C.–250 B.C.) who is known for two things: his belief that the earth revolves around the sun, and his work attempting to determine the sizes and distances of the sun and moon. Archimedes interpreted that Aristarchus meant that the ratio of the size of the earth to the size of the universe is comparable to that of the orbit of the earth compared to the sphere of stars. The number that Archimedes gave was huge!

"One can also show that the diameter of the universe is less than a line equal to a myriad diameters of the earth and that, moreover, the diameter of the universe is less than a line equal to one hundred myriad myriad stadia. As soon as one has accepted the fact that the diameter of the sun is not greater than thirty moon diameters and that the diameter of the earth is greater than the diameter of the moon, it is clear that the diameter of the sun is less than thirty diameters of the earth."

"...this number is the eighth of the eight numbers, which is one thousand myriads of eight numbers.... It is therefore obvious that the number of grains of sand filling a sphere of the size that Aristarchus lends to the sphere of fixed stars is less than one thousand myriad myriad eighth numbers."

Archimedes estimated that 8×10^{63} grains of sand fill the universe!

Tuesday — December 24, 1793 (Nivôse, du Quartidi, année II)

No church bells announced the midnight hour. No rousing peal of bells resonating in the starry night heralded the birth of Jesus. In the new Republic we are devoid of spiritual morals but we are required to worship a goddess of Reason. Notre-Dame Cathedral was turned into the Temple of Reason and a statue of the goddess of reason replaced the one of the Virgin Mary. All churches in France are now closed. No more priests, no more Christian religion, just the worshiping of reason.

Despite all that we had a most pleasant evening at home. My sister and her husband came with the baby. Madame de Maillard and several other friends of the family joined us for supper. Soon my sisters began to play the piano and the soirée became a joyful celebration. Even Papa sang and read a beautiful piece from the *De rerum natura (On the Nature of Things)* by Lucretius. This is a first century B.C. epic poem by the Roman poet and philosopher Lucretius with the goal of explaining Epicurean philosophy to a Roman audience. And then we all waited anxiously for my mother to sing, as her exquisite arias are so full of emotion.

Maman cried when she was singing an old chant de Noël: *Veni, Clavis Davidica; Regna reclude celica; Fac iter tutum superum, Et claude vias inferum.* Her eyes were wet with tears that glistened as they gently fell on her cheeks, but Maman continued with a breaking voice, *Veni, veni, Adonai, Qui populo in Sina Legem dedisiti vertice, In maiestate glorie.* We all felt like crying with her. Ours were tears of happiness.

Tuesday — December 31, 1793

The terrible year of 1793 unfolded as a nightmare that oppresses the mind with a paralyzing fear so intense that the body cannot move and one cannot speak or wake up. How can one put into context the atrocities that happened? Starting with the execution of the King of France, the Convention took drastic measures with horrific consequences; it instituted a Revolutionary Tribunal on the 9th of March, and on on the 6th of April the Committee of Public Safety was established, with formidable powers to accuse and condemn.

After that, the Reign of Terror began. Marie-Antoinette was the first victim. But many more innocent people, including aristocrats and workers, priests and nuns, educated and not educated, all have ended at the scaffold. The Revolutionary Tribunal ordered the execution of hundreds of prominent and not-so-prominent citizens.

I cannot recall all the events that changed the face of Paris on this year alone. On 1 March the émigrés were sentenced to perpetual banishment and their property confiscated. On March 18 it was decreed that any émigré or deported priest arrested on French soil should be executed within twenty-four hours. The Convention passed the Law of Suspects on September 17, which authorized the imprisonment of almost anyone and as a consequence of which hundreds were executed. As Papa says, informing has become a trade. Even some of France's most distinguished scholars have been accused and condemned. M. Condorcet is hiding from the persecution. And in November, Lavoisier, a man of vast scientific achievements who has so greatly helped the foundation of the new Republic, was also imprisoned.

What happened to the ideals of the Revolution? The promises of the Declaration of the Rights of Man have been forgotten. Hopes and dreams for freedom have turned into palpable terror. I close this page of my diary with an ancient poem that seems to reflect this reality.

> This terror, then, this darkness of the mind,
> Not sunrise with its flaring spokes of light,
> Nor glittering arrows of morning can disperse,
> But only Nature's aspect and her law,
> Which, teaching us, hath this exordium:
> Nothing from nothing ever yet was born.
> Fear holds dominion over mortality
> Only because, seeing in land and sky
> So much the cause whereof no wise they know,
> Men think Divinities are working there.
> —Lucretius, "Substance is Eternal"

Paris, France

1794

7

Knocking on Heaven's Door

Sunday — January 12, 1794

I am most happy when I am in my own metaphysical space of solitude and sanctity, disconnected from daily realities, a place where I endeavor and reside. I dream here, too. But most of all I am in heaven when I am in this special realm where new fascinating discoveries are illuminated by the bright light of knowledge.

For the past two weeks I've researched a most intriguing topic. In one of Euler's memoirs I found the term *calculi variationum*, which I understood to be different from integral or differential calculus. Because the memoir is written in Latin, I began to translate it slowly and soon it became clear that this is another branch of mathematics that deals solely with problems of maxima and minima of definite integrals. I am not sure if I translated it correctly but I will refer to this branch of mathematics as "variational calculus."

Because translation is tedious, sometimes I find solace reading about well-known isoperimetric problems, those where one seeks to answer "Among all closed curves in the plane of fixed perimeter, which curve (if any) maximizes the area of its enclosed region?" This is equivalent to the problem: Among all closed curves in the plane enclosing a fixed area, which curve (if any) minimizes the perimeter?

Isoperimetrics, an area of study that examines the largest area surrounded by a fixed perimeter, can be considered the roots for the development of the variational methodology, that Euler addressed starting with the observation made by Aristotle that most motion appears to be in either straight lines or circles. Aristotle suggested a common generating principle, namely minimization. That is, the attainment of the minimum possible value of a certain quantity such as the distance between two points.

The origin of isoperimetrics is lost in the beginning of the history of mathematics. It is known, however, that ancient Greek geometers also knew the shortest distance between two points on a sphere, and that the sphere encloses the largest volume. Mathematicians of antiquity treated the isoperimetric properties of the circle and the sphere, the last of which can be formulated in a simple form: among all bodies of the same volume, the sphere has the least boundary area.

The first proof of the isoperimetric property of the circle is due to Zenodorus (after Archimedes) who wrote a treatise on isoperimetric figures, known through the fifth book of the *Mathematical Collection by Pappus of Alexandria*. Zenodorus proved that among polygons enclosing a given area, the regular ones have the least possible length. This property implies the isoperimetric property of the circle by a standard approximation argument. Since then many proofs have been given, some of them incomplete, although employing interesting ideas.

Monday — January 13, 1794

Pappus of Alexandria solved isoperimetric problems such as this: of all plane figures with the same perimeter, which has the greatest area? He used geometrical arguments to conclude that if one considers an arbitrary closed curve in the plane of a fixed perimeter, it is the circle that encloses the largest area. The ancient Greeks were aware of this, and many civilizations have intuitively used the shape of a circle to gain more space. This led to the formulation of the isoperimetric theorem that states: "Among all shapes with an equal area, the circle will be characterized by the smallest perimeter," which is equivalent to this: "Among all shapes with equal perimeter, the circle will be characterized by the largest area."

Pappus studied the art of natural optimization. He believed that the elegant shape of the honeycomb was the result of the efficiency of Nature itself, rather than of the bee's innate sense of geometric beauty. He thought that the repeating pattern of six-sided figures seen in the cross-section of a honeycomb used the least amount of wax to build the walls. Pappus wrote this idea in an essay entitled "The Sagacity of Bees."

The sage man observed that bees keep their honey in clean and pure ways, and that they divide their combs into hexagons. Why hexagons? Pappus asked, why don't bees make each cell triangular, or square, or in some other shape? Why do the cells have straight sides in the first place? The warm wax could easily be formed into curved walls. Using straight sides, only equilateral triangles, squares and hexagons can be fitted together to leave no gaps. Pappus explained that bees "have contrived this by virtue of certain geometrical forethought, their figures must be contiguous to one another, their sides common, so that no foreign matter could enter the interstices between them, so they would not defile the purity of their produce." Pappus observed that only three regular polygons are capable by themselves of exactly filling up the space about the same point: the square, the equilateral triangle, and the hexagon. Of these, the hexagon has a larger area with the same perimeter, so "the bees conclude that the hexagon will hold more honey for the same expenditure of wax used for the perimeters of the honeycomb."

Thus, this mathematical problem is one of two-dimensional geometry. Its solution requires finding the two-dimensional shape that can be repeated endlessly to cover a large flat area, for which the total length of all the cell perimeters is the least; in other words, finding the total area of the honeycomb walls that is as small as possible.

Well, if the sphere has the smallest surface area for a volume, is it possible that hexagons with curved walls might be even more efficient for the honeycomb cells? Sure, in terms of a single honeycomb cell, if a wall bulges out we could store more honey in it for the same wall area than if the wall were straight. However, when all the cells are packed together, we can imagine how a wall that bulges out for one cell bulges in for the adjacent cell and so on. Could there be an entire honeycomb of cells with bulging walls where the net increase in efficiency of the outward bulges outweighs the net decrease caused by the inward bulges? If there were such a pattern, the honeycomb conjecture of Pappus would be false.

Pythagoras declared: "The most beautiful solid is the sphere, and the most beautiful plane figure the circle." In addition to noting the isoperimetric property of the circle (it encloses the largest area among all other isoperimetric figures), ancient geometers also noted that the sphere has the property of enclosing the largest volume among all objects with the same surface area. This property of maximal capacity formed the basis of the idea that the circle and sphere are the "embodiments of geometric perfection." In fact, this notion has been shared by most of the greatest scholars in history. Copernicus wrote, "In the first place we must observe that the universe is spherical. This is either because that figure is the most perfect, as not being articulated, but whole and complete in itself; or because it is the most capacious and therefore best suited for that which is to contain and preserve all things."

Next I will study Kepler who, in addition to discovering planetary laws, also worked on isoperimetric problems.

Thursday — February 13, 1794

Johannes Kepler (1571–1630) — the great astronomer who discovered that the planets travel on elliptical paths — conjectured that the most efficient way to pack spheres in three-dimensional space (i.e., the way that wastes the least space) was the familiar way that grocers stack oranges.

Another popular and much easier problem is Kepler's wine barrel puzzle. It requires finding the largest volume of a barrel of given diameter. The problem goes back to 1612 when Kepler was about to remarry and ordered several barrels of wine for the celebration. The salesman brought the barrels of wine and explained to Kepler how they were priced: a rod was inserted into the barrel diagonally through a small hole in the top. When the rod was removed, the length of the rod which was wet determined the price for the barrel.

Kepler was puzzled and later he wrote: "Like a bridegroom, I thought it proper to investigate the laws of measurement so useful in housekeeping." In simple terms, Kepler hoped to calculate the greatest volume of wine for a given price, and he considered different shapes for the barrels. He approximated the barrel by a cylinder. Then he asked: If the diameter is fixed, what value of the height gives the largest volume? That was before calculus, so Kepler found a formula using a method of indivisibles that defined a barrel of definite proportions.

Isoperimetrics is charmingly illustrated with the legend of the Princess Dido, that mathematical story from Virgil's *Aeneid* that I learned years ago. Stated mathematically, Dido's problem is this: *Find the optimal form of a lot of land of the maximum area S for a given perimeter L.* Clearly, the solution is a circle. Pappus, and later Theon of Alexandria, recorded that result and gave credit to Zenodorus.

Wednesday — March 19, 1794

I bought a beautiful book written by Euler in 1744. It is titled, *Methodus inveniendi lineas curvas maximi minimive proprietate gaudentes, sive so-*

lutio problematis isoperimetrici lattissimo sensu accepti, which I translated as *A method for finding curved lines enjoying properties of maximum or minimum, or solution of isoperimetric problems in the broadest accepted sense.*

This work deals with the calculus of variations, including a listing of one hundred special problems that Euler considered to illustrate his methods. He also demonstrated a general procedure for writing down the differential equation or first necessary condition. Euler also presented and discussed the principle of least action, explaining how the calculus of variations comes into play in physics.

Euler formulated the principle of least action as follows: Let the mass of the projected body be M, let v be half the square of the velocity of the projected body, and let the element of arc length along the prescribed path be ds. Among all curves passing through the same end points, the desired one makes the integral $\int M ds v^{1/2}$ a minimum, or, for constant M, $\int ds v^{1/2}$ a minimum. This principle applies to any number of bodies or particles, but it seems to run into a difficulty when one considers the motion in a resisting medium.

In the *Additamentum* II, subtitled "On the determination of the movement of projectiles in nonresistant media, by the method of maximums and minimums," I found the formulation of a general principle of minimum action (not just for the light), that is applied to the study of the trajectories described under the influence of central forces. Euler wrote: "Since all the effects of nature obey a law of maximum or minimum, there is no doubt that also in the curved trajectories that the projectiles describe ...there is some property that becomes maxima or minima." Euler added: "Then, I say that the trajectory described by the body has to be arranged in such a way that $dM s v$, or if M is constant, it has to be a minimum between all trajectories contained between such extreme".

This is a rather difficult topic that I have yet to grasp. There is so much for me to learn!

Monday – March 31, 1794

It has taken me some effort to understand the difference between integral calculus and what Euler calls "*calculi variationum*." First, it seems that the most basic problem in this branch of mathematics requires finding the function $y = y(x)$ among a class of functions that renders a given definite integral a maximum or minimum. The definite integral is of the form $\int_a^b F(x, y, y')dx$, where $y' = \frac{dy}{dx}$.

The simplest example of this type of problem is to find the shortest curve joining two points in the plane. Of course the answer is a straight line connecting the two points, something I know intuitively. However, in order to solve other not so obvious problems, I shall attempt to solve it using the methods of the calculus of variations I learned from Euler.

A necessary condition that must be satisfied by the function is the following equation that Euler derived in a memoir dated 1741:

$$\frac{d}{dx}\left(\frac{\partial F}{\partial y'}\right) - \frac{\partial F}{\partial y} = 0.$$

I could modify the basic problem by demanding that the class of functions also satisfies a side condition of the form

$$\int_a^b g(x, y, y')dx = \text{const.}$$

And thus I obtain the isoperimetric problem, such as that considered since ancient times: to find, among all closed curves of a given length, the one that encloses the maximum area.

Now let's find the shortest path in the xy-plane connecting two different points a and b: Let (x_0, y_0) and (x_1, y_1), $x_0 < x_1$, be the Cartesian coordinates of the points a and b, and $y(x)$ a function that can be differentiated twice, $x_0 \leq x \leq x_1$, such that $y(x_i) = y_i$ for $i = 0, 1$. Then ds describes a path connecting a and b, and the length of the path L is

$$L = \int_a^b ds = \int_{x_0}^{x_1} \sqrt{dx^2 + dy^2}.$$

Using $dy = y'(x)dx$ I get

$$L[y] = \int_{x_0}^{x_1} \sqrt{1 + y'(x)^2}\,dx.$$

Basically, this is the definite integral with $F(x, y, y') = \sqrt{1 + y'^2}$. Now, to minimize the integral, I solve Euler's equation:

$$\frac{d}{dx}\left(\frac{\partial F}{\partial y'}\right) - \frac{\partial F}{\partial y} = 0$$

where for this problem I have

$$\frac{\partial F}{\partial y} = 0$$

and

$$\frac{\partial F}{\partial y'} = \frac{y'}{\sqrt{1+y'^2}}.$$

Therefore, the Euler equation gives

$$\frac{d}{dx}\left(\frac{y'}{\sqrt{1+y'^2}}\right) = 0.$$

Finally, integrating with respect to x, I get

$$\frac{y'}{\sqrt{1+y'^2}} = \text{const},$$

which is the same as $y' = \text{const}$. Thus, the slope of $y(x)$ is constant, indicating that $y(x)$ is a straight line, as I already knew.

There are many other examples of maxima and minima I can attempt to solve with this method. In fact, I should try to solve the isoperimetric problems that I read about several years ago. For example, find the shape of the curve down which a bead sliding from rest and accelerated by gravity (g) will slip (without friction) from one point to another in the least time. This is the famous brachistochrone problem that was solved by Leibniz, L'Hôpital, Newton, and the Bernoulli brothers almost a hundred years ago.

For the brachistochrone problem the time t to travel from a point a to another point b is given by the integral

$$t = \int_a^b \frac{ds}{v}$$

where v is the speed, $v = \sqrt{2gy}$, and ds is the arc length as I defined before,

$$ds = \sqrt{dx^2 + dy^2} = \sqrt{1+y'^2}dx.$$

Thus I write,

$$t = \int_a^b \frac{\sqrt{1+y'^2}}{\sqrt{2gy}}dx.$$

The function to be varied is $F = (1+y'^2)^{1/2}(2gy)^{-1/2}$. Then I proceed to use the equation of Euler, which is simplified since x does not appear in my function F.

This part of the solution I leave for tomorrow, as it is almost morning and I must sleep a little before Maman notices in my eyes that I stayed all night studying. And tomorrow is my birthday, so she is probably planning a dinner party with the family.

Tuesday — April 1, 1794
(Germinal, du Duodi, année II)

I was 13 in 1789, the year I knew I'd devote my life to mathematics. Is this just a coincidence that these two are prime numbers? The next prime number is 1801. I just know something extraordinary will happen to me in the year 1801, something so wonderful I cannot yet imagine...

Saturday — April 5, 1794
(Germinal, du Sextidi, année II)

At twilight today, the most charismatic and perhaps most visionary politician of the French Revolution mounted the scaffold at the Place de la Révolution. Georges Danton has died at the guillotine, his voice silenced forever. Many admired Danton, including my father, and some credited him with saving Paris from military disorder and social anarchy in the past few years. Others implicate him in the horrific September Massacres — and it may well be that neither view is mistaken.

In the end, Danton was destroyed by his sometime ally, Robespierre. Danton had recently returned from semi-retirement on his farm to engage in this losing power struggle. Danton's earthy, all-too-human *joie de vivre* — as Papa said — was diametrically opposed to the cold-blooded Robespierre, "the Incorruptible." Danton and Robespierre came face to face as rivals for the leadership of Paris. Of the two, perhaps Danton is the easiest to sympathize with, or maybe I feel this way because of his tragic end. Of course I never met either one. I only know them through what my parents have said.

Tuesday — April 8, 1794

M. Condorcet is dead! After months of living in hiding, yesterday he was arrested in Calamart and was imprisoned in the Bourg-l'Égalité (formerly the Bourg-l'Reine). He was found dead in his prison cell this morning.

Papa came this evening and took me to his study to deliver the tragic news. I didn't cry. I just sat there mute, unable to move and listened to his words full of regret. Monsieur Condorcet hid for eight months in the house of Mme. Vernet, on rue Servandoni, just across from the Jardin du Luxembourg. M. Condorcet must have believed that he was no longer safe, because on the morning of the fifth of April he wrote his last will and left.

The same day, at about three o'clock in the afternoon, M. Condorcet arrived extremely fatigued at the door of a country house, occupied by a family who for nearly twenty years had received from Condorcet many favors. What happened then, accounts do not agree. It is believed that Condorcet solicited hospitality for a night. Something must have prevented his friends from granting his request and instead they arranged that a small garden door opening outward toward the country should not be closed at night, and that Condorcet might present himself there at ten o'clock. Papa believes that these friends undoubtedly committed the irreparable fault of delegating to others, not seeing themselves that the arrangement was carried out. For two dreary nights no door was opened to him. No one will ever know the anguish, the sufferings M. Condorcet must have endured on those long hours without any food or shelter.

Yesterday M. Condorcet, wounded and driven by hunger, entered an eating house in Clamart, and ordered an omelette. Unfortunately this wise and learned man did not know how many eggs a workman eats at a repast. When asked by the server how many he desired, he answered a dozen. This unusual number excited surprise, soon suspicion, which spread quickly around the town. The stranger was asked to show his passport; he had none. Pressed by questions, he called himself a carpenter, but the state of his hands contradicted the assertion. The municipal authorities were immediately informed; policemen came to arrest him, and took him to the prison on Bourg-la-Reine.

This morning, when the jailer opened the door of the dungeon in which M. Condorcet had been confined, in order that the gendarmes might conduct him to Paris, he found him dead.

The Marquis de Condorcet escaped the scaffold. Papa believes that his friend, Pierre Jean George Cabanis, gave Condorcet a poison that he carried in his ring and took when he realized there was no chance to escape. However, some of his friends believe that he may have been murdered (perhaps because he was too loved and respected to be executed). I don't know what to think. Although it doesn't really matter now whether he was killed or he took his own life. Monsieur Condorcet is no longer here on this world to speak and tell us the truth.

Was I expecting this turn of events? Yes and no. A part of me wanted to believe that somehow M. Condorcet would be spared, that foreign scholars would come to his aid and take him to England or Germany. I also imagined that, just as was done to defend M. Lagrange against deportation, his colleagues from the former Academy of Sciences would plead his case. Of course, the reality is that many innocent people have perished at the hands of the radical revolutionaries, and nobody seems to be able to help them.

Ironically, one of Condorcet's proposals, to found a school to educate scientists and engineers, has just been announced. The world will remember him for many contributions in mathematics, education, and philosophy. I will always remember this great man for much more.

Thursday — May 8, 1794

Antoine-Laurent de Lavoisier, a scientist in chemistry was taken to the guillotine today. Nobody was able to save this great scholar who spent his last days at the Conciergerie.

In addition to his scientific research, M. Lavoisier was also involved in business and politics. He was chairman of the board of the Discount Bank, and had a position as a tax collector in the *Ferme Générale*, a tax farming company, where he attempted to introduce reforms in the French monetary and taxation system to help peasants. Monsieur Lavoisier worked with other scientists and helped develop the metric system to secure uniformity of weights and measures throughout France. But despite his importance in the government and in science, Marat accused him of embezzling government funds and of adulterating tobacco in order to increase profits from the toll duty. This afternoon, Lavoisier, his father-in-law, and most of the other members of the *Ferme* were taken to the Place de la Révolution where they were guillotined in front of a mad crowd too ignorant to realize the hideous loss of a great man destined for glory.

I asked Papa whether M. Lavoisier's colleagues made any attempt to save him. Yes, of course, he replied, but his association with the tax farming counted so heavily against his chance for release that the Tribunal judges did not even read the report from the *Bureau de Consultation des Arts et des Métiers* written on his behalf. In part the report declared the summary of his scientific achievements, and it asserted that Citizen Lavoisier deserved to be enrolled among those whose work advanced human knowledge and who have done much to enhance the science and the glory of the nation. Among the members of the Bureau are Berthollet, Laplace, Lagrange, and Cousin.

I also asked my father if he thought M. Lavoisier was guilty to which he

Trial of Antoine-Laurent Lavoisier and the *Ferme Générale* — Paris, May 8, 1794.

simply responded "Lavoisier died innocent." I believe it, just as innocent as the King and Queen of France, and M. Condorcet and many others who have died at the scaffold. The sad part is that many more will follow them. It seems that reason, sense, and human principles cannot win over against the madness that now rules our new republic.

Monday — May 19, 1794

Fermat's principle states that light traveling between two given points follows a path requiring the least time. If the speed of light in a medium of index of refraction n is $v = ds/dt = c/n$, then the time of transit from point a to point b is

$$t = \int_a^b dt = c^{-1} \int_a^b n \cdot ds.$$

If n is a constant, then this also factors out of the integral, and the problem is reduced to the minimization of the path between two points, as done above. If n is different across regions, or varies with position, then the integral can be cast in this other form:

$$t = c^{-1} \int_a^b n(x, y) \sqrt{1 + y'^2} \, dx.$$

While it is clear why a solution that gives a minimum would be a valid path, it is not obvious to me why a maximum solution would be the right path for a light ray. But this is the general result: a solution that makes the integral unchanged will be a path for the light ray. Whether one gets a minimum or maximum, it all depends on the nature of the problem.

By Fermat's principle, t is stationary. If the path consists of two straight line segments with n constant over each segment, then $\int_a^b n \cdot ds$, and the problem can be solved by ordinary calculus. But if the index of refraction is not constant but rather depends on x or y, then the problem becomes more interesting. I shall attempt to find the path followed by a ray of light when n is proportional to y^{-1}.

Sunday — June 8, 1794
(Prairial 20 du Décadi, année II)

Among all the republican festivities I have witnessed, the *Fête de l'Être Suprême* seems the most bizarre. In contrast with the perfectly clear, seductively warm spring day, which otherwise would have been the day of Pentecost, this festival was a strange mixture of patriotic fervor, pompous drama, and just plain stupidity. We celebrated a supreme being!

At exactly five in the morning, a general recall was sounded throughout the city. It invited every citizen, men and women alike, to adorn their houses with the colors of liberty, either by hanging more flags, or by embellishing them with garlands of flowers and greenery.

Some time later, we had to assemble with our respective sections to await the departure signal. Men were to go unarmed, except for fourteen- to eighteen-year-old boys who had to carry sabers and guns or pikes and were to form a square battalion marching twelve across, in the middle of which the banners and flags of the armed force of each section were placed.

All female citizens were requested to dress in the colors of liberty. Mothers were asked to hold bouquets of roses in their hands, and young girls to carry baskets filled with flowers. Each section chose ten older men, ten mothers, ten girls from fifteen to twenty years of age, ten adolescents from fifteen to eighteen years of age, and ten male children below the age of eight to stand on the raised mountain in the *Champ de la Réunion*. We were not among them, I am relieved to say.

The ten mothers chosen by each section wore white dresses with a tri-colored sash across their chest from right to left. The ten girls were also dressed like the mothers and had flowers braided into their hair. The ten boys were armed with swords. Every citizen (us included) had to carry bouquets, garlands, or baskets of flowers. Angélique and Millie were happy to comply and adorned themselves with flowers and blue and red ribbons. I simply carried an oak branch.

At exactly eight in the morning an artillery salvo, fired from the Pont-Neuf, signaled the time to proceed to the National Garden. All citizens had to leave from their respective sections in two columns, each six abreast. The men and boys were on the right, while the women, girls, and children below the age of eight walked on the left.

Upon arrival at the National Garden, the columns of men lined up in the part of the garden on the side of the terrace called "the Feuillants," while the columns of women and children lined up on the side of the river terrace, and the square battalions of boys in the wide path in the center. When all the sections had arrived at the National Garden, a delegation went to the Convention to announce that everything was ready. The members of the National Convention arrived by way of the balcony of the Pavilion of Unity to the adjoining amphitheater. They were preceded by a large body of musicians, who were positioned on each side of the steps to the entrance. The event began with choral singing by thousands of volunteers, including a passionate rendition of "La Marseillaise."

The audience burst into applause when Maximilien Robespierre leapt onto the stage, smartly dressed in a sky-blue coat with red lapels. Robespierre, as the president, gave a rousing speech from the rostrum, explaining the reasons behind this solemn festival, and invited us to pay tribute to Nature's Creator. After his solemn speech, a symphony played. At the same time, the president, armed with the Flame of Truth, descended from the amphitheater and approached a giant papier-mâché figure raised on a circular basin, representing the monster Atheism.

From the middle of this monument, which the president set on fire, the figure of Wisdom appeared. Robespierre then led the procession off the stage and towards the Champ de Mars; behind him came a triumphal chariot pulled by eight oxen with their horns painted gold. The chariot was followed by young girls in white dresses bearing baskets of fruit, and happy mothers with arms full of roses. After this ceremony, the president returned to the rostrum and spoke again to the people. Everybody answered him with songs and cries of joy. Of course, for weeks beforehand, music teachers held singing lessons in the streets of Paris, making sure citizens knew the words to the new Hymn to the Supreme Being. I just pretended to sing.

The party culminated at 7 this evening with the Hymn to Divinity. We were tired, hungry, and so thirsty; by the time we came home, I had forgotten the speech. But I have not forgotten that this was one party invitation we could not refuse. Even ambivalence to the event could have had dire consequences.

Since early 1793, the guillotine has been cranking away with increased efficiency every day, and thousands of so-called counter-revolutionaries are crowded into the Conciergerie prison, wondering when they will die. The victims are not only nobles; the list of prisoners at the Conciergerie includes butchers, bakers, washerwomen, and seamstresses, just anybody who may be considered "antipatriotic" can end up there. And that is why we do what we're asked, even if it requires attending a frivolous party like this one.

Monday — June 23, 1794

While studying the *Mécanique analytique* of Lagrange, I came across the name Huygens. Lagrange quoted him with such admiration and so frequently in his treatment of dynamics that I had to learn about him and his work. Christiaan Huygens (1629–1695) was a Dutch mathematician and astronomer. He developed telescopes, experimented with the pendulum clock, and carried out research in timekeeping, optics, and mechanics

Huygens also constructed the first pendulum clock with a special device to ensure that the pendulum was *isochronous* — move at equal time intervals — by forcing the pendulum to swing in an arc of a cycloid. This is accomplished by placing two evolutes of inverted cycloid arcs on each side of the pendulum's point of suspension against which the pendulum is constrained to move.

The pendulum is perhaps the most important device that scholars use to describe motion. In simplest terms, a pendulum is a weight suspended from a pivot so that it can swing freely. And due to its regular motion, a pendulum is used to keep time. Beginning around 1602, Galileo Galilei was the first to study the properties of pendulums. Galileo's interest in pendulums may have developed around 1582 when he observed the rhythmic swinging motion of a chandelier in the cathedral of Pisa.

Curiously, the period (time of swing) of a simple pendulum depends on its length, the local strength of gravity, and to a small extent on the maximum angle that the pendulum swings away from the vertical (the amplitude of the swinging motion); I believe that Galileo discovered this. Then in 1673, Huygens published his mathematical analysis of pendulums, *Horologium oscillatorium sive de motu pendulorum*, an important book quoted by Lagrange on page 234 of his *Mécanique*. Years earlier other savants had observed that pendulums are not isochronous. For example Marin Mersenne (the mathematician I encountered in my studies of prime numbers) performed experiments and discovered, in 1644, that the period of pendulums depends on their width of swing, and that wide swings take longer than narrow swings.

Huygens analyzed this phenomenon and determined the shape of the curve down which a mass will slide under the influence of gravity in the same amount of time, regardless of its starting point; this is known as the "tautochrone problem." Huygens showed that this curve is a cycloid, not the circular arc of the pendulum's bob as many had believed until then; in other words, pendulums are not isochronous.

Huygens also solved the problem posed by Mersenne of how to calculate the period of a pendulum made of an arbitrarily shaped swinging rigid body, discovering the center of oscillation and its reciprocal relationship with the pivot point. In the *Horologium*, Huygens analyzed the conical pendulum, consisting of a weight on a cord moving in a circle, using the concept of centrifugal force. This force represents the effects of inertia that arise in connection with rotation and which are experienced as an outward force away from the center of rotation. Huygens also derived the formula for the period of an ideal mathematical pendulum.

Now I wonder, what is the difference between the tautochrone and the brachistochrone?

Monday — July 28, 1794
(10 Termidor, année II)

Robespierre is no longer; he was guillotined! After several weeks of executions without trial, pushing the Terror into full throttle, Parisians had enough. Yesterday, Robespierre and others were arrested. The national guards had little difficulty in making their way to the Hôtel de Ville; Robespierre was shot in the lower jaw by a young gendarme while signing an appeal to one of the sections of Paris to take up arms for him; some people suggest that the wound was self-inflicted. After a night of agony, Robespierre was taken before the tribunal, where his identity as an outlaw was proved, and without further trial he was executed with many others of his followers.

Does this mean the end of the terror? Papa thinks so. We talked about it, about the agony of this man who caused so much anguish to others, how he must have suffered in twenty-four hours all that a mortal can suffer. Hunger must have been, too, among his pangs, for his injury prevented him from eating. He had tasted no food for at least seventeen hours. Did he then comprehend what others had suffered because of him?

I do not know what to feel. At one time I had heartily embraced the republican cause; but I desired a republic as pure as it would be great, with liberty and equality for all. Now I am confused...

Friday — August 1, 1794

I found my first childhood book in a box buried in my mother's wardrobe. The brown pages had discolored and some were torn, but my pencil scribbles were still perceptible. The *catechism*, my first instruction and learning of religion, was marred by my childish strokes, showing not letters but numbers, large and small with little lines to connect them since I thought then that numbers could transform into others if they touched each other. I remember now that when I was about five my father gave me a book of arithmetic and taught me to add and subtract.

Here I am many years later, still adding and multiplying numbers, and this time the numbers are as special and magical as they seemed to my innocent mind. Today I work with prime numbers and quadratic polynomials of the form $ax^2 + bx + c$, where a, b, and c are integer coefficients and the discriminant (an expression that gives information about the nature of the polynomial's roots) is $b^2 - 4ac$.

Euler discovered in 1772 the prime-producing polynomial $x^2 - x + 41$, stating that this formula is prime for $x = 0$ to 40 and for many values thereafter. I verified, and indeed the primes for $x = 0, 1, 2, 3, \ldots$ are $p = 41, 43, 47, 53, 61, 71, \ldots$. I stopped there, but then I checked $x = 41$ and found that the formula produces $p = 1681$, which is equal to 41×41. This is a composite prime! Euler wrote in a letter to Monsieur Daniel Bernoulli in reference to a memoir published in 1771: "*Cette progression* $41, 43, 47, 53, 61, 71, 83, 97, 113, 131$ &c. *dont le terme général est* $41x + x^2$, *est d'autant plus remarquable que les* 40 *premiers termes sont tous des nombres premiers.*" So I checked $x = 43, 47, 53$ and, indeed, Euler's formula yields the numbers $p = 1847, 2203, 2797$.

Also, $f(x) = x^2 + x + 41$ is prime for all integers $x = 0, 1, \ldots, 39$. For $x = 39$ this polynomial produces $p = 1601$, indeed a prime number. But for $x = 40, 41$, it yields $1681, 1763$, neither number a prime.

I noticed that the discriminant in Euler's and Lagrange's formulas are the same, namely $b^2 - 4ac = 1 - 4(1)(41) = -163$. It makes me wonder whether this has to do with their capacity to produce primes and if so, then I should be able to find another polynomial of discriminant -163 that would also generate primes for at least 40 integer values of x. I could do it by translating Euler's polynomial. That would be a good exercise for next Monday when I will be free to do anything I wish, undisturbed for several hours while my mother and Angélique go shopping.

Sunday — August 24, 1794

Some days I cannot remember... and other times I wake up in the middle of the night, trembling with the fear I felt when the reign of terror began. On that day Robespierre declared *terror* "the order of the day." This was the beginning of brutally persecuting anyone considered enemy of the republic. The terror has claimed thousands of lives, including those of the queen Marie-Antoinette and the illustrious philosopher and mathematician M. Condorcet.

Sunday — September 7, 1794

Silence, pure and simple, devoid of startling surprises or disturbing noises. That is what I crave the most. I wish the silence of a quiet night, punctuated only by the chimes of church bells in the distance or the singing of the wind as it lulls me to sleep. It has been a while since I felt that serenity. Now I have to bury my mind in the splendorous chaste light of mathematics and close my ears to the disturbing background that does not cease. The curtains are always drawn now, to dull the noise of carriages and snorting of horses with their hooves constantly click-clacking in the street below. In this summer night it would be good to dream and think in peace.

Friday — September 19, 1794

I often wonder about the connection that exists between mathematics and reality. How something that is observable can be described by a formula or equation? Take for example the fall of an apple from my writing table. I see the apple fall and, if I am smart enough, I can figure out a mathematical relationship between its falling velocity and the distance it moved, from table to floor.

In general, to understand fully a physical process, a scientist tries to derive the process from other more fundamental concepts. For example, in the early 1600's Johannes Kepler built a model of the solar system that he then used to predict the exact locations of the planets with a precision not known until then. How did he do it? By observation! Kepler discovered planetary laws by working with Tycho Brahe's data. Brahe was a brilliant astronomer and mentor of Kepler. Without the complete series of observations of unprecedented accuracy that Brahe made, Kepler could not have discovered that planets move in elliptical orbits.

What I am saying is that usually the first step in trying to understand a physical process is to figure out *how* it works and then determine *why* it works that way. Kepler may not have known why his laws worked. It took Isaac Newton's formulation of gravity to explain why Kepler's planetary model works. At the same time, it took Newton to uncover three fundamental ways that objects interact to derive his theory of gravity. Thus, starting with Newton's three laws of motion one can derive Kepler's model: The planets move the way they do because of gravity, and gravity works the way it does because it follows three basic laws for forces.

In order to establish (scientifically) those laws of nature, Newton had to invent a new branch of mathematics. Really! He wanted to know why things move the way they do and fall the way they do. Perhaps he started thinking about speed. For example, if I walk from one place to another and keep the same pace along the way, then my walking speed at any time is the same. That speed is just the total distance I walked divided by the time it took me to move from point a to point b.

However, if instead of keeping a steady pace I vary the speed along the way, slowing down sometimes, stopping, and speeding up at other times, then I'd determine an average speed as the total distance divided by the time it took me to move from point a to point b. Fine.

But Newton may not have been satisfied with knowing the average speed. He wanted a way to know the speed of an object at any instant along its trajectory. I can see that this would create a mathematical problem since one would have to divide the distance traveled during an instant (which would be zero) by the time it took to go that distance (again, zero). And this doesn't make any sense at all since surely at each instant along the journey the moving object has some speed, or it would not move and never arrive to its destination!

So I can imagine Newton thinking that he might be able to get the speed at each instant by starting with the average speed, and making the time interval shorter and shorter until it almost became one instant of time. Of course I don't know how Newton developed the idea of "fluxion," the term he used for the derivative of a "fluent," or continuous function. All I know is that he invented the calculus in order to explain his theory of motion, starting with his notion about speed and acceleration. Of course he found plenty of other uses for it, too.

Interestingly, at the same time in Germany, another mathematician named Gottfried Leibniz invented calculus, but since he did it independently of Newton, his mathematical approach is different and has different nomenclature even if both give the same results. It seems to me that Leibniz developed calculus as a notion of pure mathematics, while Newton developed it as a tool to define physical concepts.

Now, what is gravity? Formally, gravity is the force of attraction between objects. Gravity is what makes pieces of matter clump together into planets, moons, and stars. Gravity is what makes the planets orbit the stars — like Earth orbits our star the Sun. Also, because of gravity, if I drop something, it falls down, instead of up. Well, everybody knows that! But mathematically, what is gravity and how do I use it in an equation?

I must go back to Newton's *Principia* to review these concepts and continue working with Lagrange's *Mécanique* to understand the analysis of motion.

Sunday — 28 September, 1794

This is the best news I have heard for a long time. The *Convention Nationale* proposed the establishment of the *École centrale des travaux publics*, the engineering school which Papa and his friends had been talking about for two years. It is the same school that M. Leblanc wants his son Antoine to attend to become an engineer. The Convention has already named the first professors who will teach there: Lagrange, Prony, Monge, Fourcroy, Berthollet, Guyton de Morveau, and other *hommes supérieurs*, the best scientists and mathematicians in France. In December the school of engineering will start offering classes in the *Palais Bourbon*.

Ironically, the man who submitted a project of educational reform to the

Comité d'instruction publique, the Marquis de Condorcet, is not here to smile proudly at the result of his proposal. But how I wish he was here so that he would again argue for women's rights to education. Then perhaps this new *École* would open its door to someone like me.

Friday — October 3, 1794

The term tautochrone means "same time;" it comes from the Greek *tauto-* meaning same and *chromos* meaning time. The tautochrone (also called isochrone) curve is the curve for which the time taken by an object sliding without friction in a field of uniform gravity to its lowest point is independent of its starting point. As I discovered recently, Huygens solved the tautochrone problem, proving that the curve is a cycloid. From my study of geometry I know that a cycloid is the path traced out by a point on the circumference of a circle as the circle rolls (without slipping) along a straight line.

On the other hand, the brachistochrone curve, also known as the curve of fastest descent, is the curve between two points that is covered in the least time by a body that starts at the first point with zero speed and is constrained to move along the curve to the second point, under the action of constant gravity and assuming no friction. The word brachistochrone is also derived from the Greek *brachistos*, meaning the shortest, and *chromos*, meaning time. The curve is also a cycloid, as many mathematicians (the Bernoulli brothers, Newton, Euler, Leibniz, and l'Hôpital) proved in 1697 using different methods of solution.

Here I attempt to solve the tautochrone problem by assuming that the particle's position is parameterized by the arclength $s(t)$ from the lowest point, which is proportional to the height of the curve $y(s)$. And so, based on what I learned in Lagrange's book, my governing equation is

$$y(s) = s^2$$

whose differential form is

$$dy = 2s\,ds$$
$$dy^2 = 4s^2\,ds^2$$
$$= 4y(dx^2 + dy^2).$$

Now that I have eliminated s, I can write a solvable differential equation for

dx and dy:

$$\frac{dx}{dy} = \frac{\sqrt{1-4y}}{2\sqrt{y}}.$$

How do I integrate this equation? If I change variables, for example by letting $u = \sqrt{y}$, I get this integral

$$x = \int \sqrt{1 - 4u^2}\,du.$$

I am still unsure how I can get the equation of the cycloid from the above integral. So now I try a different method and begin with the parametric equations of the cycloid:

$$x = a(\theta - \sin\theta)$$
$$y = a(1 - \cos\theta)$$

where a is the radius of the circle that rolls along a straight line.

To check that the cycloid satisfies the tautochrone property I take the derivatives,

$$x' = a(1 - \cos\theta)$$
$$y' = a\sin\theta.$$

And thus I write

$$x'^2 + y'^2 = a^2\left[(1 - 2\cos\theta + \cos^2\theta) + \sin^2\theta\right] = 2a^2(1 - \cos\theta).$$

Recalling a relationship between time of descent, path of the particle, and gravity,

$$dt = \frac{ds}{\sqrt{2gy}} = \frac{\sqrt{dx^2 + dy^2}}{\sqrt{2gy}}$$

where I can substitute my previous relationship:

$$dt = \frac{ds}{\sqrt{2gy}} = \frac{\sqrt{2a^2(1 - \cos\theta)}\,d\theta}{\sqrt{2g\left[a(1 - \cos\theta)\right]}}$$

$$dt = \sqrt{\frac{a}{g}}\,d\theta$$

$$\int_0^T dt = \sqrt{\frac{a}{g}} \int_0^\pi d\theta.$$

And so this integral gives the time that the particle takes to travel from the top of the cycloid curve to the bottom:

$$T = \left(\frac{a}{g}\right)^{1/2} \pi.$$

Voilà! This was easier than I thought. However, I need to go back and solve the problem with the first method.

Thursday – October 30, 1794 (9 Brumaire, année III)

The Convention issued a decree for the creation of the *École Normale Supérieure*. The decree states that, "A normal school shall be established in Paris to which citizens from all parts of the Republic who are already instructed in the most useful sciences shall be summoned in order to learn, under the most skillful teachers in every area, the art of teaching." This is a superior school to train teachers and professors. Messrs Monge and Lagrange were appointed to teach mathematics. The *École Normale* is expected to open in December of this year.

It is very exciting to know there will be two new schools in Paris where the greatest mathematicians will teach, the same people that I wish to meet one day. Is this one more opportunity that opens for me?

Monday – December 29, 1794

I sit here with my stack of papers and inky fingers, watching the flame of my candle dancing rhythmically, sputtering gently. The night is unusually quiet. Perhaps the bitter cold has dampened the revolutionary fervor of the soldiers patrolling the streets, as not a single gunshot disturbs this eerie silence; life in Paris seems to cease this instant. But I know there are many like me who at this very moment sit before their desks and ponder about life's meaning, students learning from books and analysis, and erudite people discovering perhaps something magnificent about the universe.

How much my life has changed in five years! Looking at the bookshelf across my bed reminds me of the journey that I began at thirteen, a voyage of discovery into the world of mathematics. Just like the progress of science, from the ancient Greeks to today, so my own learning has taken me —

from the childish arithmetic that my father taught me when I was five to the discovery of branches that require gaining a rather formidable mathematical vocabulary, including mastery of concepts and advanced analytical skills. Ever since I found the truth of my being, the direction of my life seems more certain. Mathematics is my world and to numbers and the pursuit of scientific truth I dedicate my life.

This time as the year concludes, I wish to reflect on this other world of my intellectual pursuit and pay homage to the savants that paved the way for me, a humble student who reaches out to them. I shall start at the beginning of humanity's story, soon after language developed, as it is then that I can imagine humans began counting. My own first memory of counting is with my fingers so I imagine most people do it that way naturally. Numbers must have followed, with somebody introducing a sign to represent each group of ten.

The ancients must have developed arithmetic as soon as they invented a numerical system. Different civilizations developed their own systems, and those that included both the concept of place value and the idea of zero developed mathematics further. Geometry and algebra were born. Both branches needed each other, as a number expressed, for instance, as 2 squared $(2 \cdot 2 = 2^2)$ can also be described as the area of a square with 2 as the length of each side. Equally 2 cubed $(2 \cdot 2 \cdot 2 = 2^3)$ is the volume of a cube with 2 as the length of each dimension.

It seems that in Babylon mathematics was already advanced hundreds of years before Christ, as some ancient scholars gave examples of their geometrical and algebraic calculations. The Babylonians expressed their mathematical problems in geometrical terms, but their solutions were essentially algebraic. Egyptian mathematics was less sophisticated than that of Babylon; but an entire papyrus on the subject survives. It contains a problem I learnt when I was a child: What is the size of the heap if the heap and one seventh of the heap amount to 19?

Ah, but by the sixth century B.C. a profound change occurred in the approach to mathematics with the contributions of the Greeks. The earlier (Hellenic) period is represented by Thales, Pythagoras, Plato, and Aristotle, and by the schools associated with them. The Greeks developed mathematical thought based on logical principles.

Pythagoras is perhaps the best example of a Greek mathematician who took the science into another realm and still influences the world today. In Crotona Pythagoras founded a philosophical school based on his notion that numbers are the fundamental and unchangeable truth of the universe.

He and his followers made many discoveries through the years. Among other things Pythagoras proved that whatever the shape of a triangle, its three angles always add up to the sum of two right angles (180 degrees), the first theorem I learned. The Pythagoreans even discovered the relation between mathematics and music. The Pythagorean idea that number is the cornerstone to understanding reality inspired philosophers in the fourth and fifth centuries to develop theories in physics and metaphysics using mathematical models. These theories were to become influential in medieval and current philosophy.

Then in about 300 B.C. Euclid published the *Elements*. His most important contribution to mathematics was to gather, compile, organize, and rework the mathematical concepts of his predecessors into a consistent whole, now known as Euclidean geometry. Every mathematician from ancient times to today must have learned geometry from Euclid's *Elements*, where he compiled his thirteen books of geometrical theorems. From this book I learned that the number of prime numbers is infinite, and I followed Euclid's proof to guide my learning.

Archimedes in the third century B.C. expanded geometry and went beyond. He calculated the surface area and volume of spheres and cylinders. He wrote many treatises and he discovered the law of buoyancy. In addition to his work in mechanics, Archimedes made an estimate of π and used the exhaustion theory of Eudoxus to obtain results that foreshadowed those much later of the integral calculus.

Some time around 220 B.C. Eratosthenes, the librarian of the museum at Alexandria, mapped the stars and searched for prime numbers; he did this by a laborious process now known as the Sieve of Eratosthenes. His most significant achievement was estimating for the first time the circumference of the earth. Using geometric arguments Eratosthenes determined that the earth's circumference was about 46,000 km, an amazing estimate.

Centuries later, in about 200 A.D., Diophantus of Alexandria wrote the famous and perhaps the most important treatise in number theory, *Arithmetica*. In this wonderful book Diophantus used a special sign for minus, and adopted the letter ς for the unknown quantity. Diophantus led the way for solving a type of indeterminate algebraic equation where one seeks integer values for the unknowns.

Meanwhile, Greek mathematics spread to India, China, and Japan, and it achieved its widest influence through the Arabic transmission of Greek culture. In about A.D. 825 the most significant development was algebra, and this branch of mathematical knowledge was recorded in a book written in Baghdad by al-Khwarizmi with the title *Kitab al jabr w'al-muqabala* translated as *Book of Restoration and Reduction*. The word algebra was derived from part of the title, *al jabr*. During the Renaissance, the best work on algebra was written by Girolamo Cardano and published in 1545 with the title *Ars Magna*, the "great art."

The symbols of algebra that we know today were developed in the next century. Both plus $(+)$ and minus $(-)$ were derived from abbreviations used in Latin manuscripts. The square root sign $\sqrt{}$ is perhaps a version of r used for radix (root in Latin). The equal sign $(=)$ is attributed to an English mathematician named Robert Recorde who used it first in the year 1556. In the seventeenth century, Descartes introduced the use of x, y, and z for unknown quantities, and the convention for writing squared and cubed numbers.

At the turn of the seventeenth century, Galileo promoted the use of the scientific method, observations and experimentation, to better understand the natural world, ushering in a new era of scientific thought that changed mathematics forever. In the last two centuries, mathematicians have advanced science much more than the ancients did many centuries ago. New fertile fields of mathematics have been opened. Leibniz and Newton gave us infinitesimal calculus and later Euler introduced variational calculus.

In the late 1600s, only a few people had knowledge of the infinitesimal calculus. Newton and Leibniz as its inventors knew very well the power of calculus. The brothers Jacques and Jean Bernoulli were perhaps the first mathematicians to realize how effective infinitesimal calculus was as a tool of analysis. It is thanks to the Bernoulli brothers that other mathematicians were introduced to calculus, Jacques with his published lectures and Jean with his teaching. Several sons and grandsons of the same family also contributed to mathematics through their teaching and writings. Daniel Bernoulli, son of the elder Jean, was a very good friend of Euler.

Among the few who perfected the differential and integral calculus and taught it to others were two French mathematicians, one of whom is familiar to me through his writings. Curiously, l'Hôpital was a pupil of Jean Bernoulli, another great teacher of mathematics who, in 1691, spent a few months at l'Hôpital's house in Paris just to teach him the new calculus. Five years later, l'Hôpital published a book I have, *Analyse des infiniment petits*, the first treatise that explains the principles and use of this field of mathematics.

Leonhard Euler is perhaps the most important mathematician of this century, with so many achievements that I could not enumerate all of them.

Mathematicians admire him, stating that Euler worked in almost every area of mathematics and developed the methods of the calculus, expanding its application, effectively pushing mathematics forward into a new realm of mathematical thought. Despite the fact that Euler was half-blind for much of his life and totally blind for the last seventeen years, he retained to the end a near-legendary skill at calculation. Among other wonderful theorems he derived the elegant equation $e^{i\pi} + 1 = 0$, linking five fundamental numbers of mathematics, a sublime formula that has inspired me so much and continues captivating me today.

Here in France I know that there are savants such as Lagrange and Legendre who at this very moment must be working on new discoveries. There is no doubt that their names will be etched in the annals of mathematics alongside Newton and Euler and other great scholars too numerous to name.

This history is as yet incomplete yet illuminating. Much more remains to be revealed. There are many theorems to be proved, many new mathematical concepts to be formulated, and maybe even new branches of mathematics will grow. How can I be part of this world?

I must master the works of Euler, Fermat, Legendre, and Lagrange as I am convinced that their work is important for my education and intellectual growth. But I must also be bold and find a way to climb the steps to the ivory tower and knock on its lofty door. I just need someone to leave the door ajar, just a tiny bit, enough for me to glance inside and be bathed by its sublime light.

Author's Note

Sophie's Diary is a mathematical novel inspired by French mathematician Sophie Germain. The fictional diary presents my perspective of how a young Parisian girl could have learned mathematics in the years between 1789 and 1794. As a work of historical fiction, the chronological setting is real, drawn from the history of the French Revolution, and it describes actual historical persons, but the principal character (Sophie) is fictional, even if inspired by a real person. I attempted to capture the manners and social conditions of the people and time presented in the story, with due attention paid to period detail and fidelity. Moreover, as a mathematical novel *Sophie's Diary* includes the mathematics that the young character taught herself, interspersed with interludes from the history of mathematics

Sophie Germain is the first woman in the history of mathematics to make a substantial contribution to the proof of Fermat's Last Theorem, and to the theory of elasticity as well — all that achieved within a time span of 22 years and working mostly on her own. Overcoming prejudice and numerous obstacles to her scientific endeavors, Sophie Germain became in 1815 the first woman in the history of science to win a prize from the Paris Academy of Sciences.

From her first biographers I inferred that, around 1797, Sophie Germain assumed the name of a male student at the École Polytechnique in Paris, M. LeBlanc, and submitted her own work to Lagrange (1736–1813), one of the greatest mathematicians of the eigteenth century. At that time in Paris, women were not admitted to the École or to any other school of higher education. As reported by G. Libri[1] and H. Stupuy[2] — her first biographers — Lagrange discovered Germain by accident. Germain somehow had obtained

---

[1] Libri, G. (1803–1869), full name Count Guglielmo Libri Carucci dalla Sommaja, was an Italian mathematician who worked on mathematical physics, particularly the theory of heat. He also made contributions to number

[2] Stupuy, H. (1830–1900), was a French journalist and author who delved in diverse topics such as politics, theater, and literature. Hippolyte Stupuy is known for editing work on philosophy and a biography of Sophie Germain.

his class notes and then submitted an assignment under the assumed name of M. Le Blanc. Libri wrote in the obituary notice[3] of 1833, and Stupuy in her biography[4] of 1879, that Sophie obtained the *cahiers* (published lecture notes from the École) of several courses including Fourcroy's chemistry and Lagrange's analysis. Libri commented that it was customary that at the end of the course, professors would ask the students to submit their observations (on the course) in writing. Sophie Germain, using the name LeBlanc, sent hers to Lagrange, who "praised them and then learned the true name of the author and showed his astonishment in the most flattering terms." Neither Libri nor Stupuy say how Sophie obtained the lecture notes or explain how Lagrange discovered her true identity; they did not provide dates or details of her written observations.

Nonetheless, the discovery of such a mathematical talent in a woman must have caused a sensation in the intellectual circles in Paris. Libri wrote that, after being discovered, Germain was acquainted with "all known scientists of the time, and that some communicated their work to her and others visited her in an effort to help her."

Germain's biographers refer to citoyen Cousin,[5] professor at the College of France, who offered assistance in her study. Other academicians, such as Adrien-Marie Legendre and Gaspard Monge also took an interest in Germain. However, her education was disorganized and haphazard; the type of mathematical problems her admirers offered Germain were uneven, and her overall exposure to mathematics was thus unsystematic.[6] Instead of an orderly curriculum, she received books, lecture notes, and trivial mathematical paradoxes. From this account it may appear that Germain learnt mathematics mainly from the material that those well-meaning admirers sent her when she was already 20 or older.

Yet, no biography has explained how Sophie Germain learned mathematics before that time to become so sure of her knowledge and analytical

---

[3] *Notice sur Mlle Sophie Germain*, p. 12, appearing in Germain's posthumously published work in philosophy (Ref. Germain, S., *Considérations générales sur l'état des sciences et des lettres aux différentes époques de leur culture*, Paris, impr. De Lachevardière, 1833, in–8, 102 p.)

[4] *Étude sur la vie et les uvres de Sophie Germain*, p. 12 (in *Oeuvres philosophiques de S. Germain, suivies de pensées et de lettres inédites et précédées d'une notice sur sa vie et ses uvres* par Hippolyte Stupuy, Paris, P. Ritti, 1879, in–18, 375 p.)

[5] Jacques Antoine Joseph Cousin (1739–1800), membre de l'Institut National et professeur au Collège de France. Il wrote *Introduction à l'étude de l'Astronomie Physique* (1787) et *Traité élémentaire de l'analyse mathématique* (1797).

[6] L. Bucciarelli and N. Dworsky, *Sophie Germain: an Essay in the History of the Theory of Elasticity*. D. Reidel, Boston, 1980.

skills to submit her own work to Lagrange in the first place; such a daring act would be equivalent to acquiring class notes from an advanced university mathematics course and doing the required homework, without having attended high school! The idea of *Sophie's Diary* is an attempt to answer this question: How did she learn enough mathematics to enter — or shall we say, *push her way into* — the world of Lagrange's analysis in the first place?

There is no doubt that the real Sophie Germain was a brilliant mathematician. In order to achieve the level of mastery to tackle two of the most important mathematical challenges of her time, I believe that she had to have acquired the analytical skills early on. I conjectured that, in order to have been able to perform the supposedly brilliant mathematical analysis that she submitted to Lagrange, which occurred around 1797 (she was already 21), Sophie Germain must have learned on her own much earlier, perhaps from 1789 to 1796, when she was between the ages of 13 and 20.

In 1797, J.A.J. Cousin published the mathematics book[7] that allegedly he brought to Germain (the same year) to offer his "unconditional help in pursuit of her studies." I contend that this elementary text would have been too easy or not challenging enough for Germain, as she was already 21 years old. Besides, by then she had already responded to Lagrange's lectures in analysis, which I assume were a bit more advanced. Professor Cousin had learned about Germain precisely because Lagrange had just discovered her.

One more fact supports my assertion. An encounter with Lalande in 1797, a famous astronomer and prominent figure in Parisian intellectual society, seems to have insulted Germain's intelligence. J. Jerome de Lalande[8] is mainly known for his lecturing and writing in astronomy, including *Astronomie des dames* (1785) a popular science book for women. At the recommendation of his friend J.A.J. Cousin, the 65 year old Lalande went to visit Sophie in November 1797. During the encounter, Lalande recommended to her to read his little book for ladies. Sophie was so insulted that she never again spoke to him and refused to be associated with the popular astronomer. If she had already mastered advanced mathematics, the simplified book that he suggested must have been intellectually beneath her. I suppose that Sophie would be angered by Lalande's patronizing attitude

---

[7] A copy of Cousin's book is available here:
http://www.zvdd.de/dms/load/met/?PPN=PPN578801353.

[8] Joseph-Jérôme Lefrançais de Lalande (1732–1807), French astronomer and writer. Notable works include *Traité d'astronomie* (2 vols., 1764 enlarged. edition, 4 vols., 1771–1781; 3rd ed., 3 vols., 1792), *Histoire Céleste Française* (1801), giving the places of 47,390 stars, and *Bibliographie astronomique* (1803), with a history of astronomy from 1780 to 1802.

and his assumption that she knew so little that she needed to read his overly simplified book in astronomy.

All the published research about Sophie Germain focuses on her mathematical achievements as a grown woman. The first evidence of her advanced knowledge of mathematics is her letter to Gauss on November 21, 1804, praising his *Disquisitions arithmaticae* and sharing her own work on number theory; she was already 28. Historians have also evaluated the considerable amount of material available about her contributions to the theory of elasticity, work she did between 1809 and 1815.

On the other hand, all biographical information we have about Germain stems from the writing of two people: H. Stupuy, who never meet Germain in person, and G. Libri, who became acquainted with her in 1819 through letters where they exchanged ideas about number theory, and who met her in Paris six years later; she was already 49 years old! Libri's six-page obituary notice was kind to the deceased and somewhat embellished but without much detail, and Stupuy's biographical study was largely based on fragments of her correspondence. From that vantage of time neither biographer could provide details about the woman. Records of Germain's life, outside her work, are scant; the details of her growing up are few and, most importantly, no one has told us how Sophie Germain taught herself mathematics before she submitted her work to Lagrange.

Knowing so little about Germain's childhood, I imagined her early formative years, and considered how the teenage Sophie could have learned mathematics on her own. Again, my theory is that Sophie Germain had discovered her aptitude for mathematics when she was a little girl, and as soon as she could she began, not just to read about it but to teach herself any topic that she found in books, solving problems that attracted her the most, which were in number theory, algebra, and calculus.

In writing *Sophie's Diary*, my objective was to put in perspective how Germain could have developed the strategies to pursue her own mathematical research later on, and the resourcefulness to seek a serious professional interaction with the academicians of her time. What I wanted most of all was to put into context the environment surrounding the young Sophie Germain who, against all odds, became one of the greatest women mathematicians in history. I have always been immensely impressed and captivated just imagining how her life could have unfolded during a time of great social turmoil, growing up in an era when women were not permitted in the universities; and in spite of that she overcame prejudice, resorting to using a man's name to gain access to the sources of learning. Germain was bold, resourceful, stubborn, and a bit arrogant, but these personal traits were needed for her to become one of the most distinguished mathematicians of her time.

None of the personal events portrayed in *Sophie's Diary* pertain to the real Sophie Germain. I do not know whether in the years between 1789 and 1794 she met any of the mathematicians who appear in my fictional story. It is unlikely that Germain did. Because if she had met any of them before she was 20, Germain would not have submitted her work to Lagrange using the assumed name of M. LeBlanc. I also don't know if Germain learned the same topics included in this mathematical novel, or if she mastered number theory or the calculus when she was an adolescent.

Thus, any errors in the mathematics are my own. Some were intentional, as I wanted to present a realistic perspective of a student attempting to solve equations or to learn new mathematical concepts without a guide or teacher. As a personal diary, I thought it should not have the rigour of a textbook, but rather I attempted to portray a young woman's awkwardness, the frustration of not knowing, the eagerness to learn and the joy upon mastering a subject. I also wanted to depict the resourcefulness, the stubbornness, and the passion that any smart girl like Germain would posses at this age to study and master mathematics.

Moreover, in order to make the mathematics unambiguous, I decided to use modern mathematical notation, such as exponents, natural logarithms, and some mathematical constants. I hope that the reader well-versed in the history of mathematics will treat it as a literary technique to make the story easier to read.

Sophie Germain and her work on Fermat's Last Theorem

Fermat's Last Theorem required a mathematical proof so challenging that it stumped scholars for over 300 years. The centuries-old contest started around 1630, when the French mathematician Pierre de Fermat wrote the following note on the margin of a page in his copy of *Arithmetica*: "It is impossible for a cube to be the sum of two cubes, a fourth power to be the sum of two fourth powers, or in general for any number that is a power greater than the second to be the sum of two like powers. I have discovered a truly marvelous demonstration of this proposition that this margin is too narrow to contain." What the theorem means is that the equation $z^n = x^n + y^n$ has no non-zero integer solutions for x, y and z when $n > 2$. This became known as Fermat's Last Theorem (FLT).

The underlying problem is based on a Diophantine equation found in the classical ancient book *Arithmetica*, written by the father of algebra, Diophantus of Alexandria (c. 200–c. 298 A.D.). This is a book that Fermat, Euler, Mersenne and other scholars revered. The *Arithmetica* is a collection of problems giving solutions to both determinate equations (those with a unique solution) and indeterminate algebraic equations. The method for solving the latter is now known as Diophantine analysis in the theory of numbers.

Since Fermat's time, many mathematicians devoted much of their careers to proving his Last Theorem. In 1753, one of the greatest mathematicians of all times, Leonhard Euler, developed a partial solution for $n = 3$ and 4. After Euler, the next major step forward to probe FLT in the nineteenth century was due to Marie-Sophie Germain.

After Germain, many number theorists continued expanding the general proof, but it was not until 1994, when British-American mathematician Andrew Wiles of Princeton University developed the complete, elusive proof. It took him many years and a complex and exhaustive analysis to complete. Wiles first announced his proof of Fermat's Last Theorem in 1993, but some errors pointed out by Richard Taylor (a former student) showed that proof to be incomplete. A year later, Wiles submitted a complete proof with Taylor's collaboration. Their papers were published together in 1995 in a special volume of the *Annals of Mathematics*.

In the first decades of the 1800s, modern number theory was in its infancy. Carl Friedrich Gauss (1777–1855) was working on analytic functions, called modular forms, which turned out to be important to the new approaches to solving FLT. But Gauss did not produce the sought-after general proof. Sophie Germain did develop a partial proof of Fermat's Last Theorem, competing with other mathematicians of her time. Her achievement was recorded by Adrien-Marie Legendre (1752–1833), another great number theorist. Until recently, however, what was attributed to her is the theorem that Legendre credited to her in his 1823 memoir,[9] but some of her unpublished manuscripts reveal that her achievements went beyond what is usually known as "Sophie Germain's theorem" (e.g. Ribenboim 1999, 109–122).[10]

Her scientific labor resulted in the partial proof to Fermat's Last Theorem. Sophie demonstrated the impossibility of solving the equation $x^n + y^n = z^n$, if x, y, z are not divisible by an odd prime n. This remained the most important contribution related to Fermat's Last Theorem until the next result contributed by mathematician Kummer in 1840.

---

[9] Legendre A.-M., "*Sur quelques objets d'analyse indéterminée et particulièrement sur le théorème de Fermat,*" *Mém. Ac. R. Sc. de l'Institut de France*, 6, 1823, 1–60.

[10] Ribenboim, P., *Fermat's Last Theorem for Amateurs*, Springer-Verlag (2000).

Germain, however, never published her theorem, describing it instead in correspondence with Legendre and Gauss, as she worked mostly alone and was not a member of a school or any learned society. Germain's proof was explained by Legendre when he published his own solution for exponent $n = 5$.

For over 170 years it was assumed that Sophie was a minor player in the collaboration with Legendre. However, a recent reevaluation of her manuscripts and her correspondence with Legendre and Gauss by contemporary scholars Reinhard Laubenbacher and David Pengelley indicate otherwise.[11] To date, Laubenbacher and Pengelley provide the most comprehensive assessment of Germain's contribution to proving Fermat's Last Theorem.

Professors Laubenbacher and Pengelley completed a study of Sophie Germain's extensive manuscripts on Fermat's Last Theorem, publishing a reassessment of her work in number theory. They discovered that there is much in these manuscripts beyond the single theorem for Case 1 for which she is known from a published footnote by Legendre. They found Germain's contribution to proving Fermat's Last Theorem to be much more substantial and reveal that her achievements went beyond what is known as "Sophie Germain's theorem". In her letters she described a grand plan to prove the theorem. Furthermore, they analyzed the supporting algorithms Sophie invented for this plan and found that they are based on ideas and results discovered independently much later by others, and her methods are quite different from those used by Legendre.

Assessment of Germain's letters and manuscripts provide the evidence that many of us anticipated. In fact, independently from Laubenbacher and Pengelley's work, Andrea Del Centina[12] has also transcribed and analyzed some of Germain's manuscripts, in particular one found at the Biblioteca Moreniana (Italy) and its more polished copy at the Bibliothèque Nationale (Paris, France). Laubenbacher and Pengelley, as well as Del Centina, reached conclusions that are astonishingly similar: Sophie Germain "had a full-fledged, highly developed, sophisticated plan of attack on Fermat's Last Theorem." To get the full analysis of Germain's work on that plan and on number theory in general, the reader should consult both articles.

Based on recently uncovered evidence, Sophie Germain should be considered among the principal contributors to number theory in the first quar-

---

[11] Laubenbacher, R. and D. Pengelley, "*Voici ce que j'ai trouvé*: Sophie Germain's grand plan to prove Fermat's Last Theorem," *Historia Mathematica*, 2010, doi:10.1016/j.hm.2009.12.002.

[12] Del Centina, A., "Unpublished manuscripts of Sophie Germain and a revaluation of her work on Fermat's Last Theorem," *Archive for History of Exact Sciences*, 62, 2008, pp. 349–392.

ter of the nineteenth century. And that is only part of Germain's contributions to mathematics. When I first read Bucciarelli and Dworsky's fine book,[13] I was equally impressed to learn that Sophie Germain made an important contribution to the theory of elasticity. In my early studies of vibration or elasticity no mention was ever made of the work she did in 1815, not even when making reference to Poisson and Navier, two mathematicians who knew Germain and were very familiar with her attempt to explain the behavior of vibrating plates after Chadlin. Poisson, in particular, was her chief rival in the development of the theory of elasticity, which was the subject of a prize competition proposed by the class of mathematics and physics of the Institut de France in 1811, 1813, and 1816. Germain was the only mathematician to submit an entry. On her last attempt Sophie won the grand prize for mathematics. According to Bucciarelli and Dworsky, Germain's work "did lead to a correct equation for the elastic behavior of plates and stimulated the work of Poisson, Navier, and eventually of Cauchy."

In recent decades, Sophie Germain has emerged as the embodiment of what we all aspire to be. History can name other great female mathematicians who lived before Germain and accomplished as much, but Sophie Germain did it alone. Hypatia had her father, Theon of Alexandria to teach her; Maria Agnesi had Rampinelli and other instructors; and Émilie du Châtelet hired tutors and engaged Maupertuis and Clairaut to teach her mathematics to understand Newton. Sophie Germain did not have a teacher.

Sophie Germain is unique in many respects. As Stupuy[14] remarked in 1896, *Mlle Germain fit son entrée dans le monde au murmure favorable d'une bonne renommée, après une existence toute de travail et de réserve, elle en sortit de même, quittant une œuvre impérissable et non une gloire tapageuse.* I paraphrase: Mademoiselle Germain made her entry into the (scientific) world with the favorable whisper of a good reputation; after a grueling existence of work in solitude, she departed in the same way, leaving an imperishable work and not a boisterous glory.

---

[13] Bucciarelli, L. and N. Dworsky, *Sophie Germain: an Essay in the History of the Theory of Elasticity*, D. Reidel, Boston, 1980.

[14] Stupuy, Hte (Ed.), *Œuvres Philosophiques de Sophie Germain*, nouvelle édition, Paris (France), 1896, p. 18

Biographical Sketch

Marie-Sophie Germain
(1776–1831)

Sophie Germain was born in Paris at a house on rue Saint-Denis — the birth certificate does not provide a number — on April 1, 1776, the daughter of Ambroise-François Germain and Marie-Madeleine Gruguelu. According to H. Stupuy, her first biographer on record, Sophie's father was a wealthy silk merchant (*Marchand de soie en bottes*) with a recorded address of rue St-Denis, No. 336. This house was at the level of *la Fontaine des SS. Innocents* (Fountain of the Innocents[15]).

---

[15]The Fontaine des Innocents was created between 1546 and 1549. It was built into a wall at the intersection of rue Saint-Denis and rue au Fers (now rue Berger). In 1788 the fountain was relocated to the center of a newly created square, known as the Square des Innocents. After the relocation, a fourth side was sculpted to match the existing sides, as previously there were only three sides because the fountain was set into a wall. The fountain we see today is the result of a renovation in 1856 when the fountain was placed on top of six basins along which water cascades down. Fountain of the Innocents, 43 rue Saint-Denis, 75001 Paris.

Very little is known about Sophie's childhood, but historians define the start of the French Revolution as the time of her mathematical awakening. Germain was born during the kingdom of the tragically infamous Louis XVI and Marie-Antoinette, and she came of age during a time of social strife in Paris in those tumultuous years of insurrection against the monarchy. This was also an era of mathematical revolution. France at that time made an enormous contribution to the fund of mathematical knowledge, including the development of modern analysis and mathematical physics.

When Sophie was a teenager, seven of the greatest mathematicians of the eighteenth century worked in the French Academy of Sciences, located not far from her home: Joseph-Louis Lagrange (1736–1813), Marie Jean Antoine Nicolas de Caritat, Marquis de Condorcet (1743–1794), Gaspard Monge (1746–1818), Pierre-Simon Laplace (1749–1827), Adrian-Marie Legendre (1752–1833), Jean Baptiste Joseph Fourier (1768–1830), and Siméon Denis Poisson (1781–1840). These were the great men of science who expanded and further developed the mathematics of Newton, Leibniz, Euler, the Bernoullis, Fermat, and other distinguished mathematicians in history.

Thus, I was tempted to fill in the gaps of Sophie's earlier years with some familiarity of the knowledge available at the time. Also, born into a wealthy bourgeois family, she must have had access to books to support her study and provide the basis for her own mathematical development. References to ancient mathematicians and an acquaintance with the works of Fermat, Euler, Newton, Châtelet, and Agnesi in *Sophie's Diary* seem, therefore, amply justified.

When Germain was twelve, Lagrange published his *Mécanique analytique*, one of the greatest achievements in the analytical method of the calculus and mechanics. Lagrange's work influenced later mathematicians who made the calculus rigorous. In the years between 1789 and 1794, Lagrange, Legendre, and Laplace, among other known scholars, were developing some of their most important work at the French Académie des Sciences in Paris. These were the years I chose to focus on in Germain's life and her mathematical development.

The year 1789 is important because, with the meeting of the Estates-General on May 5, the French Revolution exploded. Sophie's house on rue Saint-Denis was geographically located in a part of the city where the brutal revolts that shook Paris seemed to be centered. Les Halles, for example, the market where hundreds of angry women demonstrated and then marched to Versailles, was just steps away from Sophie's home. The Hôtel de Ville was a six-minute walk away. This building was the stage for many violent events during the French Revolution (notably the murder of the last provost of the merchants, Jacques de Flesselles, by an angry crowd on 14 July 1789; the hanging of Foulon de Doué at the Place de Grève just across; and the coup of 9 Thermidor Year II when Robespierre was shot and arrested with his followers to be guillotined the next day).

Sophie lived almost across from La Conciergerie, known then as the "antechamber to the guillotine," the prison where Queen Marie-Antoinette and many other unfortunate people spent their last days before being beheaded. The menacing building was about 500 meters away from Sophie's home, and one can imagine it was close enough for her to bear witness to many horrid events; the street below her window could have led directly to the prison. We must remember that in the Conciergerie were confined the lowest class of criminals, as well as aristocrats and political prisoners, men and women from the highest to the lowest spheres of society.

So, 1789 stands as the pivotal year for Sophie. She was 13 when the Bastille was stormed and 17 when the reign of terror turned Paris into a city of unspeakable violence. During those years, Sophie grew both psychologically and intellectually. Surrounded by the horrible reality of that time, the very air she breathed must have been a peculiar mixture of fear, elation, idealism, and anxiety. Immersion in mathematics would have been a way to erect a wall around herself when death was so near and life was so precious.

Following the social chaos brought about by the French Revolution and the brutal reign of terror that took the lives of many, an emergency council[16] was set up in Paris. Its main decision was the creation of a new school called the *École Centrale des Travaux Publics*, which was promptly launched in December 1794. Its objective was to train engineers, both civilian and military. Four hundred students were rapidly enrolled, and "revolutionary courses" in mathematics and chemistry were the foundation. J. L. Lagrange (1736–1813) was the founding professeur of analysis and Gaspard Riche de Prony (1755–1843) of mechanics; descriptive and differential geometry were taught by Gaspard Monge (1746–1818).

In September 1795, the school's name was changed to *École Polytechnique*, presumably intended to convey the idea of a plurality of techniques. The school opened in buildings attached to the Palais Bourbon[17] overlooking the Seine River and facing the Jardin des Tuileries to the east. A *Journal*

---

[16] *Rapport à la Convention Nationale*, 28 Septembre 1794.

[17] Ivor Grattan-Guinness, "The *École Polytechnique*, 1794–1850: Differences over Educational Purpose and Teaching Practice." *The American Mathematical Monthly*, Vol. 112, No. 3, March 2005, pp. 233–250.

Polytechnique was established initially to fulfill the requirement of publishing the lecture courses. Each volume was composed of a varying number of *cahiers*. The print run was a thousand copies, sometimes more. The first lesson Lagrange gave at the *École* was on May 24, 1795.[18] He retired from teaching in 1799.

Her biographers tell us that Germain obtained the class notes and sent her observations in writing to Lagrange using the name LeBlanc. Bucciarelli and Dworsky reported that there was a male student enrolled at the École with the name Antoine-August LeBlanc.[19] In Fourcy's *Histoire de l'École Polytechnique*, I found the list of students enrolled in 1794,[20] where the name "Leblanc, Ant.-Augustin" appears, followed by the note "*Elève démiss*." This suggests that the student resigned (in 1797). Is this the M. LeBlanc that Sophie impersonated?

What is recorded in her manuscripts and biographies is that Sophie Germain took a man's name, M. LeBlanc, and sometime after 1795 she submitted her analysis to Lagrange.[21] We can argue that it was not difficult to obtain the class notes, since they were published and she could have bought them. But I wonder how Sophie managed to submit her own notes to Lagrange. Did she know LeBlanc well enough to ask that he deliver her own notes, or did she simply mail them to Lagrange, knowing that the student had left the *École*?

We may never know. In *Sophie's Diary*, it became imperative to introduce a character named Antoine-Auguste LeBlanc, the son of a friend of the family with whom Sophie has some social interaction, mainly to satisfy her eagerness to learn more mathematics since he was being tutored in the subject and could share with her some of that knowledge.

Yet, while Lagrange did not teach mechanics at the *École*,[22] his mathematical analysis must have been rigorous and difficult for the students attending his lectures. According to Grattan-Guiness, Lagrange taught only his own calculus, which was purely algebraic and based upon the Taylor se-

---

[18] Ambroise Fourcy, *Histoire de l'École Polytechnique*, Oxford University, Berlin, 1828, p. 74.

[19] In the letters trascribed by Stupuy (1879) the signature is given as "LE BLANC" (two words). Bucciareli and Dworsky, who referred to the same letters, use the spelling "LeBlanc" (one word). Laubenbacher and Pengelley also use LeBlanc.

[20] *Liste Générale par promotion d'entrée des élèves de l'École Polytechnique, Promotion de 1794*. Fourcy, p. 395.

[21] Lagrange's great treatise was the *Mécanique analytique* (1788). In this he laid down the law of virtual work, and from that one fundamental principle, by the aid of the calculus of variations, deduced the whole of mechanics, both of solids and fluids.

[22] Ivor Grattan-Guinness.

ries expansion. Lagrange's courses of 1794–1796 led to his famous *Théorie des fonctions analytiques*, which appeared in book form in 1797. His course of 1799, somewhat more advanced, appeared in three forms between 1804 and 1808, in the *Journal Polytechnique* and as a book with the title *Leçons sur le calcul des fonctions*.

Thus, if Lagrange was curious to meet this Monsieur LeBlanc, Sophie must have written exceptionally smart observations to his lectures. Eventually Lagrange discovered that M. LeBlanc was actually a young woman. Impressed by her brilliance and resourcefulness, it is assumed that Lagrange became her mathematical counselor. However, I was not able to uncover any evidence of this or find out when Lagrange first met Sophie. It seems that no letters were exchanged between Germain and Lagrange during those early years, and he did not write about her in correspondence with others or in his memoirs, as Gauss did.

It is possible that Sophie Germain developed some professional relationship with other professors at the *École* and, without attending lectures, she could have expanded her initial analysis. Stupuy reported that due to the original circumstances of her appearance in the Paris scientific community, the "approval of the famous author of the *Analytical mechanics*, her age and some other details on her beginnings, all that aroused the curiosity of many and at once Mlle Germain was in contact with all the known scientists of the time." Furthermore — Stupuy suggested —"each one requested the honor to be presented to her: some communicated their work to her, others addressed their works to her, and another even visited her." One can just imagine how overwhelmed Sophie must have been by all the attention, when all she desired was the opportunity to study at the *École* like the young men of her age! She continued her work in isolation.

Her education was, however, disorganized and haphazard, since she never received the rigorous instruction that she eagerly desired. Recognizing this, perhaps, Sophie sought out scientific advice from the famous mathematicians of her time and was courageous enough to submit her own ideas and solutions to very difficult mathematical problems.

Adrien-Marie Legendre became her mentor, colleague, and friend. Legendre was another of the eminent French mathematicians, *Examinateur* at the École between 1799 and 1816.[23] I cannot conclude how he helped her or when the relationship became a scientific collaboration, but it is clear

---

[23] One major feature of the École was the distinction made between teaching and examining; so examiners (*Examinateurs*) were also appointed. For mechanics and analysis the initial pair were C. J. Bossut (1730–1814) and P. S. Laplace (1749–1827), followed by Legendre in 1799. Ref.: Grattan-Guinness (2005).

that during her work on the theory of elasticity they had important scientific discussions through the exchange of letters, leading to her winning the *Prix de Mathématiques*, a major international contest established by the French Academy of Sciences (Institut de France). What is factual is that Legendre collaborated with Germain in number theory, years after her work with Chladni's plates, and Legendre included some of her mathematical discoveries in his memoir of 1823.[24]

Before we address that research activity, let us concentrate on the first years of the 1800s, when Sophie Germain began her work on Fermat's Last Theorem. At that time, Sophie could have become acquainted with Legendre's *Essai sur la théorie des nombres (Essay on the Theory of Numbers)*, which was published in 1798. We can safely assume that she had studied it thoroughly before she began to correspond with Gauss. Sophie was undoubtedly impressed by the work of German mathematician Carl Friedrich Gauss (1777–1855). In 1801, Gauss published his masterpiece on the theory of numbers, *Disquisitiones arithmeticae* (Arithmetical Research). Three years later, a twenty-five-year-old Sophie began to correspond with Gauss, sending him some of her own work. How did she develop the audacity to write to him? The only plausible answer is that Sophie had developed a thorough understanding of the methods presented in Gauss's dissertation. In addition, seeking acceptance as a genuine mathematician, what better way than go to the prince of mathematics himself?

Between 1804 and 1809, Sophie wrote many letters to Gauss, initially adopting the pseudonym "M. LeBlanc" because she feared Gauss would ignore her letters if he knew she was a woman. In these letters she sent proofs in number theory, and Gauss praised her ingenuity and mathematical ability.

In effect, Gauss did not know she was a woman almost his own age until after the French occupation of his hometown in Germany in 1806. Sophie contacted a French commander who was a friend of her family and asked him to inquire about the well-being of Monsieur Gauss. Eventually Gauss discovered that "M. LeBlanc" was in reality Mademoiselle Sophie Germain. The circumstances of her unmasking by Gauss are well-known but the story is worth repeating here.

In the fall of 1806, when Brunswick was taken by the French, Sophie, now 30, became worried about Gauss and his safety. She appealed to Gen-

---

[24]Legendre, A.-M., *"Recherches sur quelques objets d'analyse indéterminée, et particulièrement sur le théorème de Fermat,"* *Mém. Acad. Roy. Sci. Institut de France*, 6, 1823, pp. 1–60.

eral Pernety, who was besieging Germany. On November 27, a French officer named Chantel, chief of a battalion, went looking for Gauss and found
him in his home with his wife and child. He informed Gauss that his general, Pernety, who was busy with the encampment in Breslau, had sent him
at the instance of "Demoiselle" Sophie Germain in Paris to inquire after
Gauss and if necessary offer his protection. *"Il me parut un peu confus,"*[25]
Chantel reported to his general shortly after this scene. Gauss must have
been perplexed by this visitor, because he knew neither a General Pernety
nor a Demoiselle Germain! In all Paris — Gauss explained to the officer
— he knew only one lady, Madame Lalande, the wife of the famous astronomer.

When the officer asked further whether he wanted to send a message to
Mlle Germain, Gauss did not know what to respond and simply thanked
the officer and his general for the kind attention shown him. Not until three
months later did Gauss discover who Sophie Germain was. Early in 1807,
Sophie wrote to Gauss to admit she was indeed M. Le Blanc and explained
her concern that he would not take her seriously if he knew she was a
woman. She wrote: *...je ne vous suis pas aussi parfaitement inconnue que
vous le croyez; mais que, craignant le ridicule attaché au titre de femme
savante, j'ai autrefois emprunté le nom de M. Le Blanc pour vous écrire
et vous communiquer des notes ....* It seems that Sophie believed that an
erudite woman would be ridiculed. At the bottom of her letter she indicated
her address in Paris: *"P.S. Mon adresse est: Mlle Germain, chez son père,
rue Sainte-Croix-de-la-Bretonnerie, no. 23, à Paris."*

Gauss was truly impressed by Germain's work, sufficiently to mention
her mathematical proofs and insight in a letter he wrote to his colleague
and friend Heinrich Wilhelm Matthäus Olbers (1758–1840). Yet, the correspondence with Germain ended in 1809 when Gauss stopped replying to
her letters.

Perhaps, although admiring her mathematical work, Gauss did not consider Sophie his equal or he didn't feel responsibility to mentor her. Most
likely Gauss was too busy with his own life to think about a foreign corre-
spondent.[26] At the time he was going through major career changes and had
to deal with a family tragedy in the fall of 1809. In November 1807 Gauss
moved from Brunswick to Göttingen with his pregnant wife and their young
son to begin a new position as professor of astronomy and director of the

---

[25] Stupuy, 1896, p.67.
[26] Gauss was not familiar with the French language, (See Stupuy, 1896, p. 274). And thus it
must have taken him some effort to read Sophie's letters and answer them properly.

astronomical observatory. His intellectual interest was now in astronomy, not number theory. In fact, in the summer of 1809 Gauss published his major work in astronomy, *Theoria motus corporum coelestium in section-ibus conicis solem ambientium* (*Theory of the motion of the heavenly bodies moving about the sun in conic sections*).

Family life for Gauss was eventful during this period. In February 1808 his daughter was born, and shortly after his father died. Meanwhile, Gauss was teaching at the university, carrying out pioneering research, and overseeing the construction of a new astronomical observatory. In October 1809, Gauss lost his beloved wife, a month after giving birth to a third child. Gauss was devastated and expressed his sorrow and grief to his friend Olbers with whom he sought solace immediately after burying his dear wife Johanna.

Most likely Sophie knew nothing of this. On May 14, 1810, Sophie received a most odd letter from Jean Baptiste Joseph Delambre (1749–1822)[27] in his role as *trésorier de l'Université Impériale*,[28] containing an unusual request. Delambre stated that he had received a letter from Mr. Gauss and he was entrusted with a commission on which he (Gauss) engaged him to ask her (Sophie's) opinion. Gauss had received a medal valued at 500 francs, an award funded by astronomer Lalande, but instead of money Gauss desired to have *une belle montre à pendule* to give as a gift to his wife (to be[29]), and Gauss asked that Sophie help with the selection of the object. In October 11, 1809, Gauss lost his first wife. Distraught, the grieving single father became engaged to Mina Waldech on April 1, 1810, and the wedding took place on August 4 of the same year. Thus, we can conclude that the pendulum clock was a gift Gauss intended for his new bride. In any case, there is no record of how Sophie took this request or how she helped Delambre select the gift. One is tempted to fill the gap in that story with the conclusion that Sophie Germain may have been delighted to have that privilege, and if so then Gauss should have been grateful. He could have shown more support when she asked for validation of her work related to Fermat's Last Theorem. It appears that he did not.

Between 1809 and 1815, Germain worked exclusively on developing the mathematical theory of vibrating plates, making a substantial contribution to wave phenomena. Her work was based in part on experiments on vibrating plates introduced by the German physicist Ernst F. F. Chladni. He sprinkled sand on elastic surfaces, strummed the edges with a bow, and

---

[27] French mathematician and astronomer. In 1801, Delambre was appointed Permanent Secretary for Mathematics of the Academy of Sciences, a post he held until his death.

[28] A university founded on May 10, 1806, part of the education initiative of Napoleon .

[29] Gauss remarried a few months after losing his first wife.

noted the resulting patterns, exhibiting the so-called Chladni figures. Germain used Euler's theoretical base (waves in strings) to develop her own analysis. Bucciarelli and Dworksy address in great detail this aspect of Germain's work.

As told by Stöckmann,[30] on a February evening of the year 1809 in Paris, a carriage drove up before the Tuileries Palace, the official residence of Napoleon Bonaparte, emperor of the French people. "The passengers were the Count de la Place[31], chancellor of the Senate, La Cépède, great chancellor of the Legion of Honour, Bertholet, senator of the empire, and a Dr. Chladni from Wittenberg (Germany)." Chladni had become known for his experiments on vibrating plates, and stopped for a long stay in Paris while on a lecture tour through Europe." There he met the leading contemporary scientists in person. Apart from the three already mentioned, there were Poisson, Savart, and Biot, as well as Alexander von Humboldt, among the guests to Bonaparte's residence.

After playing some musical pieces on the Clavicylinder, Chladni demonstrated the sound figures and, as he later reported, "Napoleon showed much interest in my experiments and explanations, and asked me, as an expert in mathematical questions, to explain all topics thoroughly, so that I could not take the matter too easy. He was well informed that one is not yet able to apply a calculation to areas curved in more than one direction, and that, if one were successful in this respect, it could be useful for applications to other subjects as well."

Napoleon had an excellent knowledge of mathematics,[32] and thus was able to appreciate the need for expanding on this work. The next morning Chladni got a gratuity of 6000 francs, connected to the request to publish his book *Akustik* (Acoustics) in French. In November 1809 the book was published under the title *Traité d'Acoustique*.

On the first Monday of 1809, the Class of Mathematics and Physics of the Institut de France announced the prize competition with the following challenge: *formulate a mathematical theory of elastic surfaces and indicate just how it agrees with empirical evidence.* The deadline was set at October 1, 1811, for candidates to submit such a mathematical theory. The prize, a medal of gold valued at 3000 francs, was to be awarded to the winner at the public session the first Monday of January 1812. Legendre, Laplace,

---

[30] Stöckmann, H.-J., "Chladni meets Napoleon," *Eur. Phys. J. Special Topics*, Vol. 145, 2007, pp. 15–23.

[31] Count de la Place is the mathematician better known as Pierre-Simon de Laplace.

[32] Napoleon Bonaparte graduated from the École Militaire and was examined by Pierre-Simon Laplace, whom Napoleon later appointed to the Senate of Paris.

Lagrange, Lacroix, and Malus were elected to judge the contest.

Since the theory on the vibrations of elastic plates did not yet exist, the description of Chladni's figures was rather qualitative. Two-dimensional elasticity theory was considered too formidable for most mathematicians and, in fact, Lagrange said that the mathematical methods available were inadequate. Sophie Germain, however, accepted the challenge and spent the next two years trying to derive a theory of elasticity, basing her analysis on that of Euler. In January 1811, she communicated with Legendre to discuss her approach. It is not clear how many letters were exchanged between Legendre and Germain or what exactly he provided to help her. What is a fact is that on September 21, 1811, shortly before the deadline, Sophie submitted her anonymous memoir, this being the only entry in the contest. She had derived an equation for the elastic surfaces, but her work did not win the award.

Her historians conclude that Germain had not derived her hypothesis from physical principles, and her analysis lacked the necessary rigor owing to her deficiency in analysis and the calculus of variations. Sophie's essay did generate more interest in the topic and provided needed insight to pursue the theory. Lagrange, who was one of the judges in the contest, amended Sophie's calculations and developed an equation that could better describe Chladni's experiments.

When she submitted her entry to the competition, Germain established "a flurry of correspondence" with Legendre who was in the commission to draw up the program for the prize. The first published (undated)[33] letter from Legendre provided Sophie with clarification about some apparent mistake on her analysis that had to do with the equation $\sin 1/2\omega = 0$, for which he referred her to a paper by Euler. In the following letter, dated January 19, 1811, Legendre provided further explanation on the same topic. The correspondence continued until 1813, and it is clear that the purpose of the letters was to address her analysis of the vibrating surfaces.

Without a winner in 1812, the contest deadline was extended. Once more Sophie submitted the only entry in which she demonstrated that Lagrange's equation yielded Chladni's patterns in several cases; unfortunately, she could not give a satisfactory derivation of the equation. Nevertheless, the panel of judges, which included the best mathematicians of her time, deemed Sophie's second mathematical memoir worthy of an honorable mention.

Once again the contest was re-opened and this time Sophie (now 38) seemed to be more confident because she submitted a new mathematical

---

[33] H. Stupuy, 1896, p. 291.

essay, this time under her own name. On January 8, 1815, the class of mathematical and physical sciences of the Institute of France announced in its public meeting that Sophie Germain had won the *grand prix*, a medal of one kilogram of gold. But she did not appear at the award ceremony, to the disappointment of many who wished to meet her in person.

It is not surprising to me that Sophie Germain, a shy woman, did not go to receive the accolades and collect her prize. What should have been Sophie's greatest achievement and source of pride, however, turned into a bittersweet victory. She received a curt response from Siméon Denis Poisson, one of the judges and her chief rival on the theory of elasticity, who wrote that her analysis still contained deficiencies and lacked mathematical rigour. It has been suggested[34] that she thought the judges did not fully appreciate her work, and that the scientific community did not show the respect that seemed due to her. In fact, strangely enough, it was reported that Poisson avoided any serious discussion with Sophie and ignored her in public.

I imagine Sophie's disappointment at the unwelcoming response from the scholars she so desperately sought out. Although she was the first mathematician to attempt solving such a challenging problem, and others used her elegant analysis to derive their own results, Sophie was not taken as seriously as she deserved to be. Yet, she was obstinate and intellectually driven.

Sophie extended her research and, in 1825, she submitted a paper to a commission of the Institut de France, whose members included Poisson, Gaspard de Prony, and Laplace. Sadly, the commission ignored Sophie's essay and her work went unrecognized until 1880, when it was found among de Prony's papers.

Although Sophie Germain's explanation of Chladni's patterns was incomplete, the prize was given to her all the same since it was acknowledged that her mathematical treatise signified essential progress towards the theory of elasticity.

In fact, the problem proved to be obstinate. Navier, Cauchy, and Poisson continued research in this area for many years. The complete solution for circular plates was found by Gustav Robert Kirchhoff (1824–1887) but not before 1850, and (according to Stöckmann) still in 1891 one could read in the *Handbuch der Physik*: "As to the strict mathematical theory, only a few cases are known in which it yielded results appropriate to be universally applied to the experiment."

Interestingly, in the same session of December 26, 1815, in which So-

---

[34] Louis L. Bucciarelli and Nancy Dworsky, *Sophie Germain: An Essay in the History of the Theory of Elasticity*, Dordrecht–Boston, Mass., 1980.

phie Germain was named recipient of the *Prix extraordinaire*, the Academy established a new contest: the proof of Fermat's Last Theorem. This was a topic that had fascinated Germain for many years. However, the contest, after being confirmed in 1818, was withdrawn in 1820. Yet, it must have given Sophie Germain the impetus to continue and reinvigorate her research because she immediately contacted the German savant she had sought in her youth.

On May 12, 1819, Sophie Germain sent a letter to Gauss after almost ten years of silence. She wrote:[35] *de vous faire le remerciemens que je vous dois et aussi de vous communiquer les recherches qui m'ont occupées depuis l'époque à la quelle j'ai en l'honneur de vous écrire. Quoique j'ai travaillé pendant quelque tems à la théorie des surfaces vibrantes… je n'ai jamais cessé de penser à la théorie des nombres.* [36] She wanted to reacquaint herself with him and share her research on the theory of vibrating surfaces, emphasizing that she never ceased thinking about the theory of numbers. Sophie then admits that she had reflected about Fermat's Last Theorem for a long time, ever since she read his book *Disquisitiones arithmeticae*. Germain outlined her strategy for a general proof of FLT. Once more, in closing, she asked him for a judgment: *"Je vous aurez la plus grande obligation si vous étés assez bon pour prendre la peine de me dire ce que vous pensez de la marche que j'ai suivie."*[37]

Despite the flattery, Sophie seems to have been more devoted to her own work for she did not wait for a reply. After all, she was a mature 43-year old award winner of the most prestigious scientific competition in France. Around the same time, she corresponded on the subject of FLT with Legendre, Poinsot (Del Centina found a letter from Sophie to Poinsot dated July 2, 1819), and with the very young Italian mathematician Guglielmo Libri (1802–1869) who submitted his own memoir to the Paris Academy of Sciences in 1820.

Thus, after 1819 it is clear that Germain was devoted to her grand plan to solve FLT. This was also an epoch characterized by a more mature professional relationship with Legendre and then with Libri, a mathematician who became one of her closest friends.

Libri's *Memoria sopra la teoria dei numeri* (*Memoir on the Theory of*

---

[35] Andrea Del Centina, "Letters of Sophie Germain preserved in Florence, " *Historia Mathematica*, Vol. 32, Issue 1, 2005, pp. 60–75 (quote on p. 63).

[36] This passage of Germain's letter is as transcribed by Del Centina, who respected her spelling and grammar as much as possible, just as we do.

[37] Andrea Del Centina, "Unpublished manuscripts of Sophie Germain and a revaluation of her work on Fermat's Last Theorem," *Arch. Hist. Exact Sci.*, 2008, 62:349–392 — DOI 10.1007/s00407-007-0016-4.

Numbers), written when he was 18, was published in Florence and immediately it was translated into French for Cauchy to present it to the Paris Academy. Libri's memoir was received on January 22 of 1821 and Sophie Germain must have obtained a copy for her own research.

I should emphasize that Sophie Germain worked most diligently for many years to formulate her comprehensive program to prove Fermat's Last Theorem. To give here some perspective, it would require an exposition of the work others have done, studying and analyzing Germain's newly discovered manuscripts. The reader is thus urged to consult the exquisite work authored by R. Laubenbacher and David Pengelley in 2010.[38] Their *"Voici ce que j'ai trouvé*: Sophie Germain's grand plan to prove Fermat's Last Theorem" is perhaps the best study to date.

Joseph Fourier, another distinguished French mathematician and physicist who expressed a great admiration for Sophie, befriended this great woman of his generation — he was eight years older. Their relationship is reported in correspondence they maintained from 1820 until his death in 1830. I was not able to determine when Fourier met Germain for the first time, but it could have happened in 1820 when he was asked by Legendre to review her *Recherches sur la théorie des surfaces élastiques* (published in 1821). In the preface, Germain stated that before publishing this *mémoire*, she requested counsel from Fourier.

Jean Baptiste Joseph Fourier (1768–1830) is famous for his pioneering work on the representation of functions by trigonometric series. The son of a tailor, Fourier became a teacher of mathematics at age sixteen at the military school in Auxerre. He later joined the faculty at the *École Normale* at Paris in the year of its founding (1795) when he was twenty-seven. His teaching success soon led to the offer of the Chair of Analysis at the *École Polytechnique*, and in 1807, he was elected a member of the French Academy of Sciences.

During the French Revolution, his outspoken criticism of corruption in 1794 led to the issue of a decree demanding his arrest, and eventually Fourier would be sentenced to the guillotine.[39] He went in person to Robespierre in Paris to plead his case, but he was denied. On his return to Auxerre, where Fourier lived at that time, the Comité de Surveillance issued an order for his arrest. Eight days later he was imprisoned. However, Robespierre was executed on 28 July 1794 and Fourier was immediately released.

---

[38] Laubenbacher, R. and D. Pengelley, *"Voici ce que j'ai trouvé*: Sophie Germain's grand plan to prove Fermat's Last Theorem," *Historia Mathematica*, 2010, doi:10.1016/j.hm.2009.12.002.

[39] Grattan-Guinness, *Joseph Fourier 17681830*, MIT Press, Cambridge, MA, 1972.

In 1798, Napoleon selected Fourier for the legendary expedition to Egypt where he stayed until 1801. In November of the same year he returned to Paris and, according to Foucry, Fourier resumed his teaching at the École. But shortly after, in February of 1802, Fourier was appointed Prefect of the Department of Isère, a region in the east of France. From there he wrote his famous monograph on heat diffusion, which he presented to the Institut de France in 1807. In January 1812, Fourier won the Institut prize competition on heat diffusion, and began a third version of his work as a book, which was published in 1822 as *Théorie analytique de la chaleur*. The same year he became *secrétaire perpetual de l'Académie*. In this capacity, Fourier wrote in 1823 to invite Sophie to the meetings at the Academy.

Germain also earned the respect of many others besides Legendre and Fourier. Letters from J.B. Delambre (1749–1822), A.L. Cauchy (1789–1857), L. Poinsot (1777–1859), and L.M. Henri Navier (1785–1836) written between 1820 and 1823 suggest that Sophie was, if not part of the French *pléiade d'hommes supérieurs*, at least not ignored by the savants of her time. French physicist Jean-Baptiste Biot wrote that "Mlle Germain is probably the one of her sex who has most deeply penetrated the science of mathematics, not excepting Mme du Châtelet, *for there was no Clairaut*" (referring to Emilie Châtelet's mathematical tutor)[40].

An interesting relationship developed between Sophie Germain and the Italian mathematician Guglielmo Libri in the last years of her life. She met Libri for the first time in the spring of 1825 in Paris; she was 49 and Libri was 23. What brought them together was their mutual interest in number theory. The following paragraphs were composed from material published in two papers by Andrea Del Centina written in 2005[41] and in 2008.[42]

Based on three pages believed to be written by Germain, dated June 22, 1822, and found by Del Centina in the Moreniana Library of Florence, Sophie studied Libri's memoir as another reference of interest for her own research. It is reported that Germain and Libri were in touch from 1819 on. And thus, when he visited France, it is understandable that the two should meet. Libri was in Paris from the end of December 1824 until the middle of August 1825. Sophie met Libri on May 13, 1825, at a Thursday evening party given by François Arago (1786–1853)[43] at the observatory. The day

---

[40] *Journal de Savants*, March 1817.

[41] A. Del Centina, "Letters of Sophie Germain Preserved in Florence," *Hist. Math.* 32, 2005, pp. 60–75.

[42] A. Del Centina, "Unpublished manuscripts of Sophie Germain and a revaluation of her work on Fermat's Last Theorem," *Arch. Hist. Exact Sci.*, 62, 2008, pp. 349–392.

[43] Arago was a French physicist and astronomer who made major contributions to physics. In

after, Libri wrote to his mother: "finally yesterday night I met Mademoiselle Germain who had won the mathematical prize at the Institute some years ago. I talked with her for about two hours, she has an impressive personality." Sophie must have also liked him because she invited him to her home for lunch. Del Centina concludes that the two met several times and that their relationship quickly extended beyond mathematics.

Nevertheless, no one has found letters from Germain to Libri for the four years after September 1826, and one can conclude that Sophie Germain persevered, working alone until her death. Sophie was stricken with cancer in 1829 but, undeterred by her illness and the revolution that shook Paris again in 1830, she continued working on number theory, wrote her philosophical ideas, and refined her analysis on the curvature of vibrating surfaces.[44]

Her last letter to Libri, dated May 17, 1831 (from Del Centina, 2005, p. 15) is heart-wrenching. She writes: *Je suis malade, Monsieur et très malade, j'ai fait beaucoup d'efforts pendant votre séjour ici pour ne pas vous fermer ma porte, mais le mal est bien augmenté depuis et je ne peux plus aujourd'hui ni recevoir des visites ni m'occuper. Je suis aux prise avec d'horrible souffrances ma vie est un vrai supplice aucune saison ne peut améliorer mon sort on me dit qu'avec beaucoup de tems et des soins je pourrai retrouver quelque repos.* Sophie was suffering immensely.

In her last years Germain also outlined a philosophical composition that her nephew published posthumously in 1833 as *Considérations générale sur l'état des sciences et des lettres*.[45] In this article she wrote her ideas about the general state of sciences and literature, leaving behind beautiful statements concerning her beliefs in a creator of the universe, the soul, and the human spirit. When did Sophie Germain begin to deal with philosophy? When did she write this essay? Her first biographer stated that she wrote it during her last few months of life, when the excruciating pain of cancer prevented her from continuing her mathematical research. Furthermore, it seems that her *Pensées* (*Thoughts*) were not meant for publication. Yet, one can assume that, as imperfect as her manuscript was when her nephew discovered it, such a deep and extensive philosophical essay must have been written a long time before.

---

astronomy, Arago discovered the Sun's chromosphere. He also played a part in the discovery of Neptune by Urbain Leverrier. Arago gave popular lectures in astronomy between 1813 and 1845 and became Director of the Paris Observatory.

[44] Germain, Sophie, *Discussion sur les principes de l'analyse employés dans la solution du problème des surfaces élastiques*, Annales de physique et de chimie, 1828.

[45] Germain, Sophie, *Considérations générales sur l'état des sciences et des lettres aux différentes époques de leur culture*, Paris, impr. De Lachevardière, 1833, in-8, p. 102.

Marie-Sophie Germain died in Paris on June 27, 1831. Years later, a plaque was erected on the house where she died, granting Sophie Germain the title of philosopher and mathematician. The rather unpretentious building where Sophie lived throughout her adulthood is still standing on 13 rue de Savoi, across from the Seine River, and a few blocks from rue Saint-Denis where she grew up. The plaque on the upper right side of the arched entrance to the apartment building states: *Sophie Germain, Philosophe et Mathématicienne, née a Paris en 1776 est morte dans cette maison le 27 Juin 1831.*

When the matter of honorary degrees came up in 1837 at the centenary celebration of the University of Göttingen, Gauss regretted that Sophie Germain was no longer alive.[46] She would be exceedingly proud to be awarded an academic degree, especially at the suggestion of a great scholar whom she so admired, a mathematician she sought out as her first mentor.

I wrote this book to pay tribute to the memory of Sophie Germain, outstanding woman, philosopher, and brilliant mathematician who became immortal thanks to her work with prime numbers, her partial solution to Fermat's Last Theorem, her work in mathematical physics, and her contribution to the study of elasticity.

France has honored Germain by naming a little Parisian street[47] and a high school for girls after her. On March 1, 1882, the first high school for young girls (*École Primaire Supérieure de Jeunes Filles*) was opened with 65 students. By the 1970s, the number had increased to 1600. Up until 1888, the school was known as the School of Rue de Jouy and then, as noted in the historical book at the school, it was "named after Sophie Germain, mathematician and philosopher who had to instruct herself during a time (1776–1831) when the status of women who wanted an education was a handicap." The Lycée Sophie-Germain is located at 9 rue de Jouy in the 4ᵉ Arrondisement of Paris, a four minute walk (260 meters) from the Pont-Marie over the Seine river.

The Sophie Germain Foundation, created in 2003 under the sponsorship of the Institut de France, was designed to award an annual prize of mathematics to young researchers, on the proposal of the Academy of Sciences. In 2006, a prize in mathematics in the amount of 8,000 euros was given for scholarship assistance to the students of the Lycée Sophie Germain.

---

[46] G. W. Dunnington, *Carl Friedrich Gauss: Titan of Science* (1955), University of Chicago Press, p. 68. Republished with additions by J. Gray, and F.-E. Dohse, Mathematical Association of America (2004).

[47] Rue Sophie Germain, 75014 Paris (subway station Métro Mouton-Devernet).

Let us close this book with a statement of Sophie Germain's greatest achievement.

Germain's Theorem. For an odd prime exponent p, if there exists an auxiliary prime θ such that there are no two nonzero consecutive pth powers modulo θ, nor is p itself a pth power modulo θ, then in any solution to the Fermat equation $z^p = x^p + y^p$, one of x, y, or z must be divisible by p^2.

Sophie Germain's Theorem can be applied for many prime exponents, by producing a valid auxiliary prime, to eliminate the existence of solutions to the Fermat equation involving numbers not divisible by the exponent p. This elimination is today called Case 1 of Fermat's Last Theorem. Work on Case 1 has continued to the present. The reader should consult Laubenbacher and Pengelley[48] who provide the necessary background to understand Germain's Theorem and its application.

---

[48] R. Laubenbacher and D. Pengelley, "*Voici ce que j'ai trouvé*: Sophie Germain's grand plan to prove Fermat's Last Theorem," *Historia Mathematica*, 2010, doi:10.1016/j.hm.2009.12.002.

Marie-Sophie Germain Timeline

Achievement Areas: Number theory and mathematical physics

Major Contribution: Partial Proof of Fermat's Last Theorem and Germain prime numbers. Germain is also one of the founders of the theory of elasticity.

Timeline:

1776	Marie-Sophie Germain is born April 1 in Paris, France.
1789	She discovers her passion for mathematics after reading Archimedes's story. French Revolution begins with the fall of the Bastille.
c. 1798–99	She collects lecture notes from the École Polytechnique and submits her own review for the class of Professor Joseph-Louis Lagrange using the pseudonym "M. LeBlanc." Lagrange discovers her true identity and Sophie becomes known in the intellectual world of Paris.
1804	She writes to German mathematician Carl Friedrich Gauss concerning his essay *Disquisitiones arithmeticae* (1801) and attaches her own analysis signing the letter as "M. LeBlanc." She begins her attempt to probe Fermat's Last Theorem (FLT).
1805	Gauss writes to Olbers about LeBlanc's letter.
1807	Gauss discoveres M. LeBlanc is actually Mlle. Germain and praises her work.
1811	Sophie submits a memoir with her mathematical analysis to the Institut of France as entry to the competition, explaining the vibration patterns demonstrated by Ernest Chladni.

1813	She is awarded an honorable mention for her second entry on Chladni's vibration plates.
1816	Sophie wins the *Prix de Mathématiques* awarded by the Class of Mathematics and Physics of the Institut de France for her mathematical theory of vibrations of general curved and plane elastic surfaces.
1819	Sophie writes again to Gauss, indicating she has resumed her work on Fermat's Last Theorem. She also begins correspondence with other mathematicians involved in number theory, communicating her own research.
1819–1830	Sophie works on her grand plan to solve FLT.
1823	Legendre publishes a second edition of *Theorie des nombres* where he adds a footnote with Germain's Theorem.
1825	Sophie meets Italian mathematician Libri.
1831	Sophie Germain dies in her Paris home on June 27.

Bibliography

Works of Sophie Germain

Germain, Sophie, *Recherches sur les théories des surfaces élastiques*, Paris, Vè Courcier, 1821, in-4, avec planches.

Germain, Sophie, *Remarques sur la nature, les bornes et l'étendue de la question des surfaces élastiques et équation générale de ces surfaces*, Paris, impr. de Huzard-Courcier, 1826, in-4, 21 p.

Germain, Sophie, *Discussion sur les principes de l'analyse employés dans la solution du problème des surfaces élastiques*, Annales de physique et de chimie, 1828.

Germain, Sophie, *Mémoire sur la courbure des surfaces élastiques*, Annales de Crelle, Berlin, 1831.

Germain, Sophie, *Considérations générales sur l'état des sciences et des lettres aux différentes époques de leur culture*, Paris, impr. De Lachevardière, 1833, in-8, 102 p.

Germain, Sophie, *Oeuvres philosophiques de S. Germain, suivies de pensées et de lettres inédites et précédées d'une notice sur sa vie et ses oeuvres par Hippolyte Stupuy*, Paris, P. Ritti, 1879, in-18, 375 p.

Germain, Sophie, *Mémoire sur l'emploi de l'épaisseur dans la théorie des surfaces élastiques*, Paris, Gauthier-Villars, 1880, in-4, 64 p. (extrait du journal de mathématiques pures et appliquées, 3è série Tome 6, 1880).

Works on Sophie Germain

Biedenkapp, **G.**, *Sophie Germain, ein weiblicher Denker*, Jena, 1910

Blay, Michel, Robert Halleux, "La science classique, XVIe-XVIIIe siècle," dictionnaire critique, Flammarion, 1998.

Boncompagni B., *"Cinq lettres de Sophie Germain à Charles-Frederic Gauss,"* Arch. der Math. Phys., 1880, Vol. 63, pp. 27–31, Vol. 66, pp. 3–10.

Bucciarelli, Louis L. and Nancy Dworsky, *Sophie Germain. An essay in the history of the theory of elasticity*, D. Reidel Publishing Company, Dordrecht, Boston, London 1980.

Charpentier, Debra, "Women Mathematicians," the *Two-Year College Mathematics Journal*, Vol. 8, No. 2, Mar., 1977, p. 73–79.

Coolidge, Julian L., "Six Female Mathematicians," *Scripta Mathematica*, Vol. 17, March/June 1951, pp. 20–31.

Dahan-Dalmédico, Amy, "*Mécanique et théorie des surfaces: les travaux de Sophie Germain*," *Historia Math*. Vol. 4, 1987, pp. 347–365.

Dahan-Dalmédico, Amy, *Sophie Germain*, Pour la science, dossier hors série janvier 1994, Pour la science No. 132, Octobre 1988.

Dahan-Dalmédico, Amy, *Aspects de la mathématisation au XIXè siècle. Entre physique et mathématique du continu et mécanique moléculaire, la voie d'A-L Cauchy*, Thèse Sciences, Nantes, 1990 (disponible à la bibliothèque universitaire de lettres sciences-humaines de Nantes).

Dahan-Dalmédico, Amy, "Sophie Germain," *Scientific American* 265, 1991, pp. 117–122.

Dahan-Dalmédico, Amy, *Sophie Germain*, in *Spektrum der Wissenschaft* 2, 1992, pp. 80–87.

Dahan-Dalmédico, Amy, *Mathematisations: Augustin-Louis Cauchy et l'Ecole Française*, Paris, Editions du Choix, 1992.

Deakin, Michael, "Women in mathematics: fact versus fabulation," *Austral. Math. Soc. Gaz.* Vol. 19, 1992, p. 105–114.

Del Centina, Andrea, "Unpublished manuscripts of Sophie Germain and a revaluation of her work on Fermat's Last Theorem," *Arch. Hist. Exact Sci.* Vol. 62, 2008, pp. 349–392.

Del Centina, Andrea, "Letters of Sophie Germain preserved in Florence," *Hist. Math.* 32, 2005, pp. 60–75.

Deledicq, André, *Histoires de maths : K. F. Gauss, S. Ramanujan, Sophie Germain, E. Galois et un tableau chronologique de mathématiciens*, Paris, Berger Levrault, Art-Culture-Lecture, 1992.

Deledicq, André and Dominique Izoard, *Histoires de maths*, ACL — Les éditions du Kangourou, 1998, pp. 39–41.

Dickson, L.E., *History of the Theory of Numbers*, New York, 1950.

Dubner, H., "Large Sophie Germain Primes," *Math. Comput.* 65, 1996, pp. 393-396.

Edwards, H., *The last theorem of Fermat*, Mir, Moscow, 1980.

Eves, Howard, *An Introduction to the History of Mathematics* (chapter 13), Saunders Series, Saunders College Publishing, Philadelphia PA, 1990.

Ford, D. and V. Jha, "On Wendt's Determinant and Sophie Germain's Theorem," *Experimental Math.* Vol. 2 , 1993, pp. 113119.

Franklin, Christine Ladd, "Sophie Germain: An Unknown Mathematician," Century, Vol. 48, 1894, pp. 946–949, [Reprinted in the AWM Newsletter, Vol. 11, No. 3, 1981, p. 711].

Friedelmeyer, Jean-Pierre, *L'histoire des mathématiques par correspondance*, pp. 57–61, L'Ouvert 88, 1997.

Gauss, K.F., "Letter from Gauss to Sophie Germain, 30 April 1807," In *Zur Geschichte der Theorie der kubischen und biquadratischen Reste*, In Gauss: Oeuvres Complètes, Vol.XI pp.70–74.

Genocchi, Angelo, Realis, S., "*Inforno ad una propozione inesalta di Sofia Germain*," *Bolletino di Bibliografia e di storia delle scienze mathematiche e fisiche*, Vol.17, 1884, pp. 315–316.

Genocchi, Angelo, "*Teoremi di Sofia Germain inforno a i residui biquadratici*," *Bolletino di Bibliografia e di storia delle scienze mathematiche e fisiche*, Vol.17, 1884, p.248–251.

Gianni Micheli, "The philosophical works of Sophie Germain," p. 712–729 in *Scienza e filosofia, Saggi in onore di Ludovico Geymonat* (Ed.: Corrado Mangione) Garzanti Editore S.p.A., Milan, 1985, p. 860.

Grattan-Guiness, I., "Recent researches in French mathematical physics of the early 19th century," *Ann. of Sci.* Vol. 38, no.6, 1981, pp. 663–690.

Gray, Mary W., "Sophie Germain: a bicentennial appreciation," *AWM Newsletter*, Vol.6, No.6, Sept.–Oct. 1976, pp. 10–14.

Gray, Mary W., "Sophie Germain," p. 47–56 in *Women of Mathematics. A biobibliographic sourcebook*, Eds.: Louise S. Grinstein and Paul J. Campbell, Greenwood Press Inc., Westport, Conn., 1987.

Gunther, Siegmund, "*Il carteggio tra Gauss e Sofia Germain*," *Bolletino di Bibliografia e di storia delle scienze mathematiche e fisiche*, vol.15 (1882), p.174–175. Also in Zeitsch. *Math. Phys.*, Hist.–Lit. Abt., Vol. 26, 1881, pp.19–25.

Hauchecorne, Bertrand, Surateau, Daniel, Des mathématiciens de A à Z, Ellipses, 1996.

Haven K., *Marvels of science: 50 fascinating 5-minute read*, Libraries Unlimited, 1996. Grade 3 & up. p. 238.

Heydemann, Marie-Claude, *Histoire de quelques mathématiciennes*, Publications mathématiques d'Orsay No. 86–74.16, in P. Samuel: Mathématiques, Mathématiciens et Société, 1974.

Hill, A.M., Sophie Germain: A Mathematical Biography. A B.A. thesis presented to the Department of Mathematics and the Honors College of the University of Oregon, August 1995.

Indlekofer, K. H., and A. Járai, "Largest Known Twin Primes and Sophie Germain Primes." *Math. Comput.* 68, 1317–1324, 1999.

Jaroszewska, Magdalena, "Portraits of women mathematicians," p. 23–29 in Report on the fifth annual EWM meeting, CIRM, Luminy, France, December 9–13, 1991.

Kelley L., "Why were so few mathematicians female?," *Mathematics Teacher*, Vol. 89, No. 7, Oct. 1996, pp. 592–596.

Klens, Ulrike, *Mathematikerinnen im 18. Jahrhundert: Maria G. Agnesi, G.-E. du Châtelet, Sophie Germain. Fallstudien zur Wechselwirkung von Wissenschaft und Philosophie im Zeitalter der Aufklaerung.* Forum Frauengeschichte, Bd. 12, Pfaffenweiler, 1994.

Krasner, M., *A propos du critère de Sophie Germain*— Furtwängler pour le premier cas du théorème de Fermat, Mathematica Cluj. 16, 1940, pp. 109–114.

Ladd-Franklin, Christine, "Sophie Germain: an unknown mathematician," *Century*, Vol.48, Oct. 1894, Reprinted in AWM Newsletter, Vol.11 No.3, May-June 1981, pp.7–11.

Laubenbacher, Reinhard and David Pengelley, *"Voici ce que j'ai trouvé*: Sophie Germain's grand plan to prove Fermat's Last Theorem," *Historia Mathematica*, Vol. 37, 2010, pp. 641–92.

Martinez, J.A.F., "Sophie Germain," *Sci. Mon.*, Vol. 63, 1946, pp. 257–260.

Micheli, G., "The philosophical works of Sophie Germain," *Scienza e filosofia*, Milan, 1985, pp. 712–729.

Mordell, L. J., *Three Lectures on Fermat's Last Theorem*, Cambridge University Press, 1921, p. 30.

Osen, Lynn M., *Women in Mathematics*, Massachusetts Institute of Technology Press, Cambridge MA, London, 1974.

Perl, Teri, *Math Equals. Biographies of Women Mathematicians* + Related Activities, Addison Wesley, Menlo Park, 1978.

Petrovich, V.C., "Women and the Paris Academy of Sciences," *Eighteenth-Century Studies*, Vol. 32, No. 3, Constructions of Femininity, Spring, 1999, pp. 383–390.

Rashed, R., *Sciences à l'époque de la Révolution Française.* Recherches historiques, Librairie du Bicentenaire de la Révolution Française. Librairie Scientifique et Technique Albert Blanchard, Paris, 1988, p. 474.

Sampson, J.H., "Sophie Germain and the theory of numbers," *Arch. Hist. Exact Sci.* 41, 1990, No. 2, pp. 157–161.

Smith, Sanderson and Greer Lleaud, *Sophie Germain, Notable Women in Mathematics: A Biographical Dictionary*, Charlene Morrow and Teri Perl, Editors, Greenwood Press, 1998, pp. 62–66.

Simalarides, A., "Sophie Germain's Principle and Lucas numbers," *Math. Scand.* 67, 1990, pp. 167–176.

Singal, Asha Rani, "Women mathematicians of the past: some observations," *Math. Ed.* Vol. 3, No. 1, 1986, , pp. 9–18.

Stupuy, H., *Notice sur la vie et les oeuvres de Sophie Germain*, Oeuvres philosophique de Sophie Germain, Paris, 1879, pp.1–92.

Tee, G. J., "The pioneering women mathematicians," *The Mathematical Intelligencer* 5, 1983, pp. 27–36.

Terquem, M., "Sophie Germain," *Bulletin de bibliographie, d'histoire et de biographie mathématiques*, Vol. 6, 1860, pp. 9–13.

Thomas, M. and A. Kempis, "An Appreciation of Sophie Germain," *National Mathematics Magazine*, Vol. 14, No. 2, Nov. 1939, pp. 81–90.

Truesdell, Clifford, "Sophie Germain: fame earned by stubborn error," *Boll. Storia. Sci. Mat.* Vol. 11, No. 2, 1991, pp. 3–24.

Vella, D. and A. Vella, "Cycles in the Generalized Fibonacci Modulo a Prime," *Mathematics Magazine*, Vol. 75, No. 4, Oct., 2002, pp. 294–299.

Waterhouse, W. C., "A counterexample for Germain," *American Mathematical Monthly*, 101, 1994, pp. 140–150.

Wussing, Hans and Wolfgang Arnold, *Biographien bedeutender Mathematiker*, 4ed Volk und Wissen, Volkseigener Verlag, Berlin, 1989.

List of Illustrations

Page 110: *The arrival of the royal family in Paris after their failed attempt to flee the country, June 25, 1791* — The Pierpont Morgan Library, New York. PML 140205 #54. Bequest; Gordon N. Ray; 1987.

Page 151: *The storming of the Tuileries Palace, August 10, 1792* — The Pierpont Morgan Library, New York. PML 140205 #67. Bequest; Gordon N. Ray; 1987.

Page 152: *Attack on the Palais-Royal* — Engraving from *A Pictorial History of the World's Great Nations from the Earliest Dates to the Present Time*, by Charlotte M. Yonge, published by Selmar Hesse, NY, 1882.

Page 153: *The royal family is taken to be imprisoned in the Temple, August 13, 1792* — The Pierpont Morgan Library, New York. PML 140205 #69. Bequest; Gordon N. Ray; 1987.

Page 156: *The massacres of the prisoners, September 2–5, 1792* — The Pierpont Morgan Library, New York. PML 140205 #72. Bequest; Gordon N. Ray; 1987.

Page 207: *First Fête de la Raison (Festival of Reason) at Notre-Dame, November 10, 1793 — Estampe de la collection Hennin*, Bibliothèque Nationale (Paris).

Page 225: *Trial of Antoine-Laurent Lavoisier and the Ferme Générale – Paris, May 8, 1794* — The Oesper Collections, University of Cincinnati (UC). Image in L. Fuguier, Vies des savants, Vol. 5, Hachette: Paris (1874). Image courtesy of Professor William Jensen from UC.

Page 249: *Sophie Germain*, a sketch from H. Stupuy, reproduced from L.L. Bucciarelli's book.

Full citation for PML 140205

Tableaux de la révolution française, ou, Collection de quarante-huit gravures représentant les événements principaux qui ont eu lieu en France depuis …le 20 juin 1789 …Ces gravures, fruit des veilles d'une société d'artistes, seront accompag,.
Paris: Briffault de la Charprais & Madame l'Esclarpet, [1791–1796]
The Pierpont Morgan Library, New York. PML 140205. Bequest; Gordon N. Ray; 1987.

Acknowledgements

This book would not have been published without the support and involvement of many people. In particular I wish to express my deepest gratitude to Professor David Pengelley, from New Mexico State University. David wrote a kind review of *Sophie's Diary* in the *Mathematical Intelligencer*, and strongly encouraged me to publish this new edition.

I also wish to express my most heartfelt thanks to Don Albers, Editorial Director of the Mathematical Association of America (MAA), for believing in this project.

I am especially indebted to Gerald L. Alexanderson, Valeriote Professor of Science at Santa Clara University and Editor for the Spectrum Series at MAA. I thank Jerry for his thoughtful editing, and for his help preparing this work for publication. Jerry's encouragement and excellent advice to improve the manuscript made this journey very rewarding.

To the members of the Spectrum Editorial Board I must express my earnest thankfulness, for reading the manuscript and for providing me with many helpful suggestions and insightful comments. And to Carol Baxter, Managing Editor at MAA, to the copyeditor who meticulously proofread every page, and to Beverly Ruedi for her expertise in preparing this book for publication, I am most thankful. Of course, any mistakes and inconsistencies found in the text are my own, and I take full responsibility.

I thank the following people and organizations for permission to use the engravings to illustrate this book: Professor L.L. Bucciarelli, The Pierpont Morgan Library in New York, Professor W. Jensen from the University of Cincinnati; and a very grateful acknowledgement to Dr. Lukas Gloor who kindly provided me with the art work for the cover.

Finally, and most sincerely, I wish to thank my beautiful and accomplished daughters Dasi and Lauren, and my husband and best friend Zdzislaw for their loving support. Lauren deserves a special recognition. My young daughter proofread my first, very rough draft, and pointed out my grammatical errors, suggesting ways to improve the text. I dedicate this book to my beloved family.

Index